Prescriptions, Patients, Profits, Perils and Pro-People Choices

Also by the Author

Pharmaceutical Sector Resources in the United States with Expertise in
International Health

The African-American Pharmacists' Socioeconomic Empowerpedia

Self-Management Diabetes Care Manual

How to Save Thousands on Prescription Medication

Prescriptions, Patients, Profits, Perils and Pro-People Choices

Be Drug Smart, Slash Costs, Be Safer, Be Healthier

Ira Charles Robinson, Ph.D.

International Pharmaceutical Health Care Expert

iUniverse, Inc.

New York Lincoln Shanghai

Prescriptions, Patients, Profits, Perils and Pro-People Choices
Be Drug Smart, Slash Costs, Be Safer, Be Healthier

iUniverse books may be ordered through booksellers or by contacting:

iUniverse
2021 Pine Lake Road, Suite 100
Lincoln, NE 68512
www.iuniverse.com
1-800-Authors (1-800-288-4677)

Because of the dynamic nature of the Internet, any Web addresses or links contained in this book may have changed since publication and may no longer be valid.

The information, ideas, and suggestions in this book are not intended as a substitute for professional medical advice. Before following any suggestions contained in this book, you should consult your personal physician. Neither the author nor the publisher shall be liable or responsible for any loss or damage allegedly arising as a consequence of your use or application of any information or suggestions in this book.

ISBN: 978-0-595-43817-4 (pbk)
ISBN: 978-0-595-88146-8 (ebk)

Printed in the United States of America

Contents

List of Tables

Preface

All too often, the high cost of prescription medication is touted as if it is the most important consumer health problem. Admittedly, prescription drug costs are a major problem for many. Because drug prices can be crucial in determining consumer access to life-saving and life-enhancing prescription drugs, I made it a priority to research and discuss as much useful cost-related, pricing-related and other appropriate drug-related information in this book. In doing so, however, the matter of drug safety must always be a primary consideration among the many vital issues affecting the use of drug therapy, and it figures prominently in all discussions throughout this book.

Consequently, I researched and waded through literally thousands of pages of Federal government regulatory documents, enforcement reports, and fact sheets from the U.S. Food and Drug Administration, the Centers for Disease Control and Prevention, and other agencies in the National Institutes of Health, the Federal Trade Commission, the Institute of Medicine of the National Academies, and the Department of Justice; drug trend and industry reports and fact sheets from the chain and independent drug store industry, the drug wholesale industry, the managed care industry, the major brand-name prescription drug industry, generic pharmaceutical industry, non-prescription drug industry, dietary and herbal supplement industries, and the homeopathic industry.

Further, I obtained and scanned pertinent congressional testimonies, press releases, research articles in medical and pharmaceutical journals, and various types of information from consumer advocacy and drug safety organizations; as well as innumerable direct email communications from various trade and professional groups as well as leading national and international research organizations. This was done to give readers as much "insider" or otherwise difficult to find information, as would be useful and appropriate, in order to ensure the currency, comprehensiveness, and accuracy of the information in this book.

In considering the prevalence, costs and consequences of some of the most serious patient drug safety issues, the following statistics point up several additional medication-related issues which affect both consumer safety and overall drug and health care costs:

1. *Adverse drug effects* affect an estimated 2.2 million patients in hospitals alone and some 350,000 nursing home patients annually, resulting in serious injuries, longer hospital stays, billions of dollars in treatment costs, and thousands of deaths.

2. *Medication errors*—At least 1.5 million people are sickened, injured or killed due to medication errors in U.S. hospitals annually; and treatment of drug-related injuries in hospitals alone cost some $3.5 billion every year.

3. *Antibiotic resistance*—More than 50 million unneeded antibiotic prescriptions are written each year for patients outside of hospitals and some 2 million patients get infections while in a hospital, of which more than 90,000 die due to antibiotic resistance. Antibiotic resistance is a global problem of epidemic proportions.

4. *Patient/consumer non-compliance*—Some 125,000 deaths result annually from patients' failure to follow instructions for their prescribed medication; further, about 10 percent of hospital admissions and 23 percent of nursing home admissions are caused by consumer drug non-compliance.

5. Callous *prescription drug misuse, overuse and abuse* promotes a drug culture leading to increasing recreational use of these potent products. Added to abuse of illegal substances, alcohol and tobacco, this further damages consumers' health.

6. *Consumer literacy*—Roughly half of all patients make medication errors and less literate consumers are more likely to make them,

with costs to the national economy estimated to be between $5 billion to $9 billion annually.

It is never acceptable to see anyone suffer physically, mentally or economically due to hindered access to needed prescription medications or due to any adverse consequences of using these in health care. This is especially so when patients suffer needlessly from a host of largely preventable medication-related problems, industrial conspiracies, and governmental lapses in enforcement of laws designed to protect consumer safety and access to affordable drug therapy.

Although many medication-related problems are inherent in the nature of drugs or otherwise caused primarily by health care system failures, an astounding reality is that drug consumers themselves are increasingly complicit in worsening these problems, as well as unwittingly causing other such problems, to their own detriment, sometimes paying the ultimate penalty—death.

What, then, is a common denominator in all of this for the consumer? Lack of information, misinformation, and a consumer's inability or simple failure to read and use readily available drug-related and health care information, make up a big component of the great failures in drug therapy outcomes in the United States.

Education and information, then, are two of the most vital keys to empowering consumers to take greater charge of their health care and partner with their health care professionals in better guarding their health. Education is the single most effective intervention for empowering and shielding drug consumers from the ravages of these potentially fatal but largely preventable problems—whether health provider-sponsored, community-based, or self-initiated.

It is for this reason that I am writing this book. In doing so, I committed myself to drawing upon the full breadth of my own personal experiences as a patient/consumer, as well as my considerable professional experience in pharmacy practice, business and the pharmaceutical industry, to introduce the consumer to a broad base of drug-related knowledge as a basis for understanding the major factors affecting the consumer drug experience—both positively and negatively—in these United States. This book is the result of that commitment, but it is merely a first step in the direction of empowering motivated, health conscious consumers to improve their health while reducing their drug-related and overall health care expenses.

In writing this book, I was able to capitalize on my vast experience over four and a half decades of developing new drug products as a research scientist, inventor and research administrator at the world's largest brand-name pharmaceutical manufacturer; educating and training pharmacists as a professor and dean of two of the nation's leading colleges of pharmacy; consulting with physicians nationwide on appropriate drug therapy for patients as a managed care pharmacist and pharmacist newsletter editor at the nation's largest pharmacy benefits manager; filling, dispensing and counseling patients on countless thousands of prescriptions as a registered pharmacist; as well as owning several business enterprises.

My entrepreneurial experience has included owning and operating small businesses in retail, wholesale and generic pharmaceutical manufacturing as well as a private practice providing disease self-management training for patients with chronic diseases like diabetes. Further, it includes my activities as an international pharmaceutical consultant on essential drug selection, procurement, manufacturing, use and evaluation, and medical supply logistics management on country-wide and multi-country projects funded by the U.S. Department of State, the U.S. Agency for International Development, the World Bank, the American Public Health Association, and non-government organizations and enterprises. These consulting and technology transfer activities enabled me to help bring increased quantities of quality essential drugs to countries in several regions of the world and improved personnel training materials to drug-handling personnel in some 70 countries.

Behind each of the countless tens of thousands of prescriptions I've filled and dispensed is a patient whose continued existence was often in my hands and who depended on me to provide rapid relief and improved quality of life from pain and suffering caused by an infection or other disease condition, and whose health and well-being impacted the lives and livelihoods of many others, including family members, other loved ones, associates and clients.

For each prescription, pharmacists must use their highly specialized knowledge of drug use, composition, therapeutic and adverse effects, as well as legal mandates and restrictions, insurance and cost-sharing, costs and pricing, social and cultural behaviors, literacy and communication, to provide the individually beneficial healing or palliative products and services ordered by prescribers. In doing so, pharmacists review each patient's on-going medications to prevent harmful drug interactions with other drugs, foods and dietary supplements, and patient age, gender and disease conditions, and drug allergies, as well as to prevent over-dosing, to bring about the positive therapeutic outcomes each drug is designed to give that patient.

Most physicians, pharmacists and registered nurses are sensitive to the needs of their patients and will also consider factors which may impede patients' access to the medications they need to stay alive, to enable them to return to work to provide for themselves and their families, and to improve their quality of life and longevity.

This book is the first of its kind and is the first in a series of books designed to assist health-conscious, self-motivated individuals at various educational levels to become drug-smart consumers. The goal is to help consumers become better informed about drug therapy and health matters and motivate them to commit to life-long learning in order to get the most from prescribed and self-selected drugs safely, effectively, efficiently and economically. The ultimate objective is to empower the average consumer who uses prescription and OTC drugs, as well as dietary and herbal supplements, to become a more informed, proactive drug-smart consumer.

The need for elevating health literacy and drug literacy at various levels in this country is undeniably great. However, it will take additional books to begin to do justice to that task for the most vulnerable segments of our society. I became convinced along the way that it would not be possible to accomplish that objective in this book. However, I fully intend to seek out collaborative partners to start on the road to accomplishing that objective, focusing on drug literacy, in additional books in this series in the near future.

I am indebted to many who have encouraged me to make this journey as well as those who agreed to review the manuscript from the perspectives of consumer, secondary school educator, pharmacist, physician, nurse, and public health educator. The reviewers do not necessarily agree with the views expressed in this book whether their critique applies to the entire manuscript or sections of it, which I was free to use as deemed appropriate in improving the book. While the assistance of reviewers is extremely valuable in refining this work, I alone am responsible for any errors or other shortcomings in this book.

These reviewers include, but are not limited to, Rondall Allen, Pharm. D., Capt. Edward A. Basdekian (USN), R.Ph., M.H.S.; Lillie R. Biddle, M.Ed.; Andrew Henry, M.D. (endocrinologist); Inez B. Joseph, Ph.D., R.N.; Douglas L. Keene, R.Ph., Pharm.D., M.H.S.; William L. Robinson, Jr., M.Ed.; Andrew Walker, M.H.S. (public health educator); and A. Sylvia Webster, D.O.; as well as my four adult children, two of whom are pharmacists and two of whom are physicians (one of whom is both a pharmacist and physician), whose names are omitted to protect their privacy.

Needless to say, I am indebted to my wife, Clarice, and other family members who tolerated the mountains of research materials, reports, books, other papers and files, and my extended self-imposed exile while planning, researching, organizing, writing, rewriting, updating, and otherwise creating the manuscript which culminated in this important book.

Ira Charles Robinson, Ph.D.
Tampa Bay Area, Florida

Chapter 1: Introduction

Our nation has become a society of "pill poppers" who rely too heavily on drugs and other chemical substances for our bodies and minds. A culture of pill popping promises dire consequences for a people victimized by its own gullibility, unhealthy lifestyles, self-destructive drug use behaviors, and drug literacy challenges. Widespread prescription and non-prescription drug overuse, misuse and abuse, and other types of chemical dependency among the young and the old are all-too-familiar symptoms of such a culture. Indeed, ours has become a culture of instant gratification and self-aggrandizement in great need of healing in more ways than one.

It is important to understand at the outset that this is not a book designed to discourage consumers from using pharmaceutical drugs. Rather it is a deliberate attempt to better educate motivated drug consumers, and professionals who routinely communicate health information to potential drug consumers, and provide them with additional tools to promote more rational, safer and more economical drug therapy. Hopefully, this book will also help in dispelling myths such as equating level of education to competence in health literacy, although education is the basic foundation upon which a more health literate citizenry must be built.

The author seeks to assure consumers of the value of FDA-approved drugs and overall safety of most when used properly by those who need them. It definitely is not an attempt to scare anyone away from using properly prescribed medication or, for that matter, responsibly selected non-prescription medication or dietary and herbal supplements, as needed. Consumers who need drugs for medically useful purposes, as determined by their prescribing doctors, should feel confident in using them so long as they comply with the verbal, written and printed instructions accompanying these products, and so long as they are willing to contribute to their own safety and positive outcomes by learning more about their own health conditions and drug effects on their bodies.

Individual prescription drug prices and the costs of prescription drugs as an integral component of overall health care costs in the United States remain a major concern. The good news is that annual rates of price increases for prescription drugs have made a dramatic turnaround within the past few years. Prescription drug costs now account for about 10 cents of every dollar spent for health care in the United States, compared to almost 20 cents just a few years ago.

Increasing expenditures for overall health care costs, including prescription costs, assume special significance for highly vulnerable groups like the elderly and the disabled, many of whom live on fixed incomes, and the uninsured and under-insured low- and middle-income families. These increases are associated with a variety of consumer-relevant issues such as a profit-driven pharmaceutical manufacturing industry, deficiencies in the nation's federal government regulatory and legislative machinery, inefficiencies in health care delivery, and medication-related problems in America's health care system, consumer literacy deficiencies, and widespread instances of modifiable self-destructive consumer behavior in the use of legitimate prescription and non-prescription drugs.

Drug use patterns and practices in America are a lesson in paradoxes. On the one hand, drug sales statistics, coupled with research data on self-destructive behaviors associated with drug use, point to clear evidence of potentially excessive use of drugs. On the other hand, research data point to a major disparity between levels of consumer prescription drug purchases and actual drug consumption. This is further exemplified by widespread consumer failure to comply with instructions for taking their medications, especially with respect to taking the full course of prescribed drugs for chronic conditions or antibiotics to prevent and treat infections.

These paradoxes strongly suggest that consumers know too little about these powerful products, and especially the political and economic driving forces behind the discovery, development, production, marketing and promotion of these products. There are also issues about the consumer's level of awareness of forces which propel expensive drug products into existence and simultaneously create artificially high demand for such drug products, specifically those still under patent protection and which are the most expensive. Strong evidence points to a major failure of our educational and health care systems, the pharmaceutical manufacturers, and our political leadership to better educate our consumers to prepare them for more successful use of information on the true nature and limitations of drugs in health care.

Too many consumers are unable to read, understand and apply drug-related information in ways to greatly reduce their exposure to serious medication-related problems, or to reduce their prescription drug costs significantly, improve their drug therapy outcomes, or trim their overall health care costs, even within the existing health care and drug pricing systems. Although not nearly as widespread or severe, the results among the more highly educated consumers are too frequently similar, in large part due to callous attitudes about the risks involved in taking drugs, a false sense of their own superior health literacy, and a stubborn reluctance to change unhealthy behaviors which have been proven to lead to debilitating disease and suffering.

Add to this world of multi-cultural literacy challenges a widespread consumer reluctance to adopt healthier lifestyles as well as an equally widespread failure of consumers to follow written and verbal instructions for taking and handling prescribed and non-prescription medications. Together these make for a recipe for certain disaster.

This book starts with some of the basics and proceeds chapter-wise through a process to expand the knowledge base of the motivated drug consumer, as well as health and education professionals, as a foundation for improved drug literacy by improving their patient communications, without becoming overly technical. The scope of information ranges from basic chemical and physical characteristics of drugs; various classifications of drugs; dosage form information important for the proper use and handling of prescription and non-prescription drugs; basic types and mechanisms of drug actions in the body; the risks and consequences associated with improper use, storage, and handling of these products; to the most basic, inherent risks associated with the use of any drug.

In succeeding chapters, the reader is given a basic understanding of where new drugs come from—physically and conceptually—and the major processes, the politics and the magnitude of costs involved in their discovery, clinical studies, product research and development, safety evaluation, regulatory approval, production and marketing of these therapeutic miracles. These are written with a view toward enhancing consumer understanding of the relationships between major brand or single-source, patent-protected drugs, and the more economical generic alternatives. A distinction between prescription and non-prescription drugs, as well as between drugs and dietary and herbal supplements, are aspects of consumer literacy which can vastly affect and influence consumer health practices, their treatment outcomes, their quality of life and, yes, their out-of-pocket costs.

The reader is taken on a journey from the pharmaceutical manufacturer's warehouse through the drug wholesaler to the dispensing pharmacy to provide a better understanding of how prescription drug costs are determined as well as to illustrate the basic level of safety and the potential for entry into our national drug supply chain of fake or substandard drugs. Further discussions introduce the reader to a plethora of political and market forces, and business tactics and strategies which can and do significantly adversely impact consumer interests during the approval process, promotion, distribution and sale of drugs.

Two chapters on medication errors and medication-related problems cover the spectrum from those caused by or contributed to by health care professionals and our health care system, to those which more knowledgeable, motivated consumers can do a great deal to reduce or prevent, by making simple changes in their selection, purchasing and use of drugs.

Finally, the reader is introduced to a second great potential alternative to the higher priced brand-name prescription drugs for highly motivated, responsible consumers who self-educate as well as self-medicate. The potential and limitations of the self-care alternative are discussed along with the role of the more economical and responsible use of non-prescription drugs and dietary and herbal supplements in health care. A limited discussion of complementary and alternative medicines is included in Chapter 9, in conjunction with discussions of over-the-counter medication in self-care activities which can be valuable in improving and maintaining good health while economizing on health care costs.

In each chapter, pressing drug consumer-related issues, and a discussion of their significance to the average consumer are highlighted with emphasis provided through estimated costs and consequences of these to the consumer in health, safety and economic terms.

Throughout this book, the reader will be introduced to a multitude of unfamiliar terms and abbreviations along with many new concepts used in health care and the pharmaceutical industry in particular. These terms, abbreviations and concepts are defined in a consumer-friendly way within the body of the text for maximum impact, understanding and retention.

In this book, the term "Big Pharma" is used to connote members of the major brand-name drug industry. Manufacturers of generic copies of Big Pharma's previously patented brand-name drugs are represented by their own trade group. Throughout this book, the word "drug" in general refers to legal drug products, whether used properly and legally or abused. Illegal drugs and other abused substances are identified as illegal or illicit drugs or substances in this book. Abbreviations in this book follow the first appearance in this book of the term or organization they represent.

Reading this book indicates that the reader has already made a life-changing decision about drugs and his or her health. Indeed, this is a significant first step toward becoming a drug smart consumer. The ultimate goal is to be able to select, obtain and use drug- and related health care information intelligently and responsibly as a primary means of protecting and improving individual and family health.

The author's concern is the result of his unusually broad experience in helping consumers of all educational, social and economic levels navigate the world of drug therapy while being impacted by mild to disabling sickness, pain and suffering of virtually every type and description. The author's attempts to make the path as smooth as possible for every consumer/patient have been invigorated by his being able to contrast the ofttimes subverted use of the nation's plenteous supply of pharmaceutical drugs in non-essential, non-productive recreational activity to that of many poor countries where essential drugs are perennially in such short supply and available drugs are beyond the affordability of the vast majority of the population.

A focus on patient-centered drug therapy and positive treatment outcomes through more deliberate medication therapy management has been central to his endeavors as a registered pharmacist of some 45 years at both of today's top two national pharmacy chains as well as at several independent pharmacies in four states and the District of Columbia; research scientist and industrial pharmacist at the nation's largest pharmaceutical manufacturer; former professor and dean at two colleges of pharmacy; manager/majority owner of pharmaceutical retail, wholesale and manufacturing operations; managed care pharmacist at the nation's largest pharmacy benefits manager, then owned by a top multi-national pharmaceutical manufacturer; diabetes educator in private practice; international pharmaceutical health care consultant on projects funded by or in collaboration with the U.S. State Department, the U.S. Agency for International Development, the U.S. Food and Drug Administration, the World Bank, the American Public Health Association, and other organizations and ventures; as well as other health-related activities.

Now the journey to and through the Pill Farm begins.

SECTION I: Pills and Portions and Their Journey to the Consumer

Chapter 2: The Anatomy and Construct of a Pill

That sparkling, innocent looking "pill" glancing up at its new owner expectantly can be disarmingly deceitful in its appearance. It can be such a beauty with its colorful, shiny coating, and artistic shape and imprint of its name or combination of numbers and alphabets to ensure its proper identification and acceptance. However, its humble appearance belies its potent power to alter vital body functions or structures or to annihilate disease-causing microbes before they have a chance of overpowering the body's immune system to wreak havoc on the body. Those who fail to respect its awesome power not only may fail to reap the full benefit of its miraculous healing potential but also may inherit the full force of its wrath.

The pill's costly and lengthy pathway from discovery to the consumer can be mind-boggling and magnificent at the same time for one curious enough to journey into its exciting past. The star performer of this production is the active new drug and its beginning is the genesis of this saga.

It landed in its well-protected, sealed bin only after having survived a complex and tortuous path of hunt and seek over some 12 to 15 years and racking up costs approaching $800 million for its short trip from the chemist's or biologist's lab to its present location. Alone the drug is powerless to do any good without undergoing a major transformation into a pharmaceutical formulation to both constrain and release its potent power as its delivery system to the human body. In order for this transformation to take place, it commanded the cooperation of a vast support staff of body builders, facilitators and offensive and defensive accomplices, all of which had to be compatible with it and each other and fully supportive of its mission. Then and only then could it begin its fateful journey to becoming the medical miracle, true to its design, to be prescribed, carefully dispensed, and finally delivered to ailing patients.

Along the way, a cadre of machinery was required—from the mighty chemical reactor to the lowly filter. Oven and spray driers, sifters, powder pulverizers, blenders and such marched into formation to sequentially, uniformly and strategically disperse the assigned helpers around the star performer. In anticipation of landing on a distant pharmacy shelf, this active drug humbly coexisted in shapeless form with millions of companions in some massive powder mix—artfully and scientifically blended with the variety of chosen helpers. These inert helpers were precisely selected and pressed into service to enable the active drug to take on a bodily form suitable for presentation to a patient for his or her acceptance, convenience and, most importantly, healing.

Following several interim processes to get to its character and shape, this pill and its millions of identical companions withstood some 12 tons of pressure per square inch as they were compressed on a rotary tablet press costing some $75,000 or, perhaps, an advanced high speed rotary tablet press with touch screen controls, costing some $150,000 or more. Their high pressure experience was, however, only momentary, because they were filled into the recesses of tablet dies, compressed, ejected and spewed out of the tablet press machine at speeds up to more than 250,000 tablets per hour (that's some 4,400 tablets per minute!).

During the tablet production process, a few of pill's devoted companions were sacrificed for testing to ensure the uniformity of the survivors as well as ensure the resulting potency of the star performer and its release from the tablet would be right on spec. Still, after all of the care taken for these assurances of quality and appearance, it suffered the indignity of standing quarantined in bulk along with its millions of companion tablets until formally released by the quality control (QC) police. The function of the QC police is to make sure that the product meets all of the specifications set by its manufacturer and approved by the U.S. Food and Drug Administration (FDA), such that the drug product will be safe to take, contain the amount of active drug required, and that this characterizes the total batch of millions.

Then, and only then, could it continue on its way through the high tech counters, bottlers, cotton stuffers, cappers, labelers, and other packaging machinery—and final quality control checks—and on to the manufacturer's warehouse. And whew! It is now ready to begin the final legs of its journey to the waiting consumer, by way of the pharmaceutical supply chain to retail or community pharmacies, hospitals and other health care institutions, and other dispensaries, either directly or, in most cases, through drug wholesale distributors.

A Pill Is a Pill Is a ... Healing Gem

Well this pill, whether it is a tablet or capsule or other pharmaceutical dosage form, has arrived in all of its splendor, miraculous healing power, and eagerness to fulfil its destiny of delivering an accurate amount of specified drug to the consumer conveniently as prescribed. It is now ready to work its healing, preventive or palliative powers on its user, provided the patient thoroughly reads and follows the prescriber's verbal instructions, directions printed on the prescription container, warnings and cautions from the pharmacist, and the printed drug information received with the prescription.

Let's assume that the patient just picked up the prescription from the local pharmacy, rushed home and counted out the one pill s/he is scheduled to take right away. The patient glanced at the front label of the prescription vial to confirm the number of pills to be taken. The label reads "Take 1 tablet by mouth 3 times daily." (In the popular jargon these are referred to as "pills", whether they are tablets or capsules or troches, because that is the term commonly used and the patient simply doesn't have the time to bother making a distinction). S/he quickly places one of these therapeutic gems on the tongue with only a casual glance and, squeezing the ever-present cell phone between head and shoulder, reaches for the nearest beverage with which to wash it down.

The patient is too tired and too busy, so s/he simply ignores the smaller cautionary or warning labels on the side of the prescription vial and relaxes as s/he gobbles her little miracle pill down the gullet with a refreshing gin and tonic. And now s/he is ready to handle more important matters. Right?

Three times daily simply means taking a pill conveniently at breakfast, lunch and dinner. Right? Or was the "pill" an antibiotic for an infection and should be taken around the clock—every 8 hours—in order to be most effective against the infectious bacteria it is designed to stymy or obliterate? Perhaps, the pharmacist tried to instruct the patient verbally as s/he snatched up the medication at the prescription counter—or perhaps the pharmacist added a printed or written note with these instructions on the prescription bag as the patient busily paid for the medication and hurriedly whisked the purchase away, still chatting away on the cell phone.

Or maybe the pharmacist instructed the patient to take the medication on an empty stomach. Before, after or with a meal? Or with a full glass of water? Or not to take it with bread or dairy products or alcohol? Or to take only as needed or until all pills are taken? Indeed, all of these instructions are aimed at making sure that the consumer achieves the positive treatment outcome the mighty pill is designed to bring about with minimal risk, when taken properly.

Taking such a casual attitude while taking prescription medication is likely to initiate a horrific cycle for the patient in spite of good intentions and the overall general safety and effectiveness of approved medicines in this country. All label and verbal instructions for taking prescription medication are important. Consumer questions to the prescriber and the pharmacist are important for clarifying when and how often to take prescribed medication, how long to take it, what kinds of problems may be encountered while taking it, what other drugs or any food not to be taken while taking the medication, and what to do if and when any bothersome or serious reactions occur.

Consider this: That innocent looking "pill" is actually a complex work of both art and science involving simple to high technology in its construction and manufacture; and it is designed to deliver an accurate dose of a potent chemical (or biological) to the body in a form where a normal digestive system and physiology will allow it to quickly disintegrate once it has been taken. This enables the active drug to dissolve in the digestive juices and the water added when it is taken. The active drug is then absorbed through the walls of the stomach and/or intestines into the blood circulating in the body. Circulating blood then carries the chemical or biological substance throughout the body to confer its miraculous benefits of healing at its appropriate site of action.

Anything that significantly interferes with this process, or otherwise adversely affects any step in this process can intensify or reduce or nullify the desired therapeutic benefits the precious drug is engineered to give. The outcome of such interference can be total treatment failure and, even worse, a worsening of the condition being treated. On the other hand, it can result in an intensification of usually mild side-effects of a drug, or even worsening of another existing health condition.

Definition of a Drug

The meaning of the word "drug" is defined in The Federal Food, Drug and Cosmetic Act. This statute defines a drug as an article intended for use in the diagnosis, cure, mitigation, treatment or prevention of disease or intended to affect any structure or any function of man or other animals.[1]

More precisely, a drug is defined by U.S. law as any substance, other than a food or device, intended to be used in the diagnosis, cure, relief, treatment or prevention of disease, or intended to affect the structure or function of the body. In this publication, when the term "drug" is used alone, it refers to a legal or federally approved pharmaceutical drug intended for legitimate medical uses and not merely for recreational or other illegal use. It refers to pharmaceuticals which require a prescription from a physician or other licensed prescriber or is available for purchase over-the-counter (OTC) without a prescription.

Drugs include natural and synthesized medicinal substances designed to prevent, treat or cure disease or other unhealthy conditions. They are mostly foreign substances introduced into the body and their use always involves an element of risk and a balancing of known risks against potential benefits. They are potent chemical or biological entities capable of affecting the functions and structure of the human body in Gram or minute milligram, or microgram doses.

All proprietary (brand-name or trademarked) drug products have at least three names: a *chemical name* which is based on rules of chemical nomenclature and can be rather long and difficult to spell or remember; a *generic name,* which is a non-proprietary name assigned by a national drug naming body in the United States, or an international drug naming body; and a *brand name* (also called the *trade name* or *proprietary name*), which is assigned by the manufacturer of the drug and registered with the U.S. Patent Office. We will revisit this observation later while discussing drug classification schemes.

The various ways active drugs are classified, irrespective of the dosage form used to deliver the drug, are discussed in some of the following sections. Drugs are available commercially in a multitude of different pharmaceutical dosage forms and are rarely taken in pure form because of their potency and the small amounts necessary to give a therapeutic action without being too toxic. The major types of pharmaceutical dosage forms, and the basic components of the most commonly used dosage form types and their functions are also discussed below, along with some of the special benefits and limitations on their use by consumers.

As we move through later chapters of this book, it will become more apparent as to why this definition is important and why consumers should take note of differences between drugs and dietary and herbal supplements. These differences determine how these products are developed and marketed, types of testing required, if any, before they are approved for marketing in the United States, and how they are regulated after initial marketing. These, in turn, relate to margins of safety to be expected during their use as well as the vast differences in their costs.

This is precisely why drugs should never be used callously as if they were mere commodities or are recreational products to be taken with carefree abandon.

The Many Faces and Parts of Pills

The current popular usage of the term "pill" to refer to pharmaceutical dosage forms is blatantly incorrect. This popular usage of the term inaccurately refers to a broad range of solid oral dosage forms which are primarily tablets and capsules but which, in fact, are differentiated by physical form, construction, and performance mechanism.

A "pill" is a long outdated dosage form generally no longer commercially available in the United States, except rarely by extemporaneous compounding by pharmacists on special order of a doctor. This old pharmaceutical dosage form was comprised of an active medicinal agent prepared in a mixture with the consistency of flour dough and then rolled into a ball shape and dried. For the most part, those products were natural products or botanicals. By far, the most popular dosage forms in use today are the tablet and the capsule. Since these and similar physical dosage forms are referred to as "pills", the term "pill" will be used intermittently in this book to facilitate understanding while at the same time expanding the reader's repertoire of more accurate drug dosage form terminology.

As indicated earlier in this chapter, a "pill" is actually a combination of the active drug substance and several inert or inactive substances required to provide a drug dosage in a convenient form to enable consumers to take accurate doses of their medication. The following discussion explains the multi-component composition of the average "pill" or, more accurately, solid oral pharmaceutical dosage form; and common types of inactive components in addition to the active drug. It also describes the contributions of each common type of inactive component.

Pills From Active Drug and Its Many Helpers

Even "simple" tablet and capsule dosage forms contain other members of the team accompanying an active drug and may involve several technologies, from simple to complex, in their formulation and manufacturing. Among the guardians and helpers in the simple "pill" the consumer receives are components like the following, each playing separate, essential functions in the processing, manufacture, release of the active drug in the body once taken by a consumer:

1. *Fillers* are inert or inactive ingredients used to provide bulk or build the body of a tablet or capsule so that it is large enough for safe and convenient handling by patients and care-givers, as well as to ensure that the product is reliably reproducible in the manufacturing plant so that it looks the same, feels the same, is the same in potency and performs the same from one production lot to the next. Fillers are the *"body builders"* in tablets and capsules.

2. *Binders* are inert ingredients to hold together the active ingredients, fillers and the other components of the tablet when compressed together on tablet presses during manufacturing. They keep tablets from chipping and breaking easily during packaging operations, during transport from the factory to warehouses and from wholesale distributor warehouses to pharmacies, as well as while sitting on pharmacy shelves for the duration of the time frame the expiration date on the label indicates for their safe use and, finally, during the handling of the finished drug product after it is dispensed to the consumer. These are members of *the defense team* which protect the cohesiveness of all members of the team.

3. *Disintegrants* or *disintegrating agents* are inert ingredients which go on the offensive once a tablet or capsule is taken by mouth. They ensure that the highly compressed, hard tablet or capsule containing powders or granules or other sophisticated components disintegrates promptly to release the active drug in a timely manner in the stomach or, when so designed, in the intestines. A disintegrating agent is a facilitator and member of the offense, just waiting to go to work at the appropriate time.

4. *Lubricants* are inactive substances which enable the powder mixtures and pre-formulated granules incorporating the active drug to flow freely and smoothly into the hollow dies in which standard pharmaceutical tablets are heavily compressed during

the manufacturing process or, in the case of capsules, the open capsule shell during filling. They also prevent the compressible powder or granular mixture from sticking to the sides of the tablet dies and the surfaces of the tablet punches which compress the materials into tablets and when the tablets are ejected from the tablet machine. These lubricants are highly effective in small quantities.

5. A *glidant* may also be used to facilitate the free and uniform flow of finely ground powder mixtures into the hollow tablet dies for tablet compression or into the empty capsule shells for uniform filling during manufacturing. Finely ground powders tend to clump together rather than flow freely and glidants prevent this clumping and are usually effective in extremely minute quantities. A glidant is a specialist used only when needed.

6. *Coatings.* Many tablets and capsules are treated with a variety of coatings, most of which are applied by a spraying and drying technique and some of which are applied by compression of powder mixtures over previously compressed tablets. Why coat these? There are several important reasons that tablets and capsules may be coated, although most of these are packaged, distributed, dispensed and taken exactly as they leave the tablet press or capsule filling machine. Coatings perform a variety of functions. Some of the functions coatings perform are to:

    a. Make a tablet or capsule easier to swallow when placed in the mouth;

    b. Mask the bitterness or other unpleasant taste of dosage units taken orally;

    c. Make the dosage unit more tamper proof or tamper evident;

    d. Make a dosage unit more difficult to counterfeit; or to

    e. Delay the release of the active drug until it reaches the intestines in order to prevent or reduce irritation of the stomach or to ensure faster or more complete absorption when released in the intestines.

7. *Coloring agents*, which must be approved for food and drug use, have two principal functions in pharmaceutical dosage forms: safety and aesthetics or appearance. In performing these important functions, they:

a. Help to ensure that a person gets the right drug and the right dose of the drug by distinguishing between different drug products of

    i. different strengths of the same drug;

    ii. different dosage formulations of the same drug in strengths corresponding to those of another formulation; and

    iii. different dosage formulations of the same drug from different sources (manufacturers) in the same dosage form available from multiple manufacturers.

b. Render the dosage units more pleasing to the eyes of consumers and health care professionals, hopefully, making the medication-taking experience more acceptable.

The above discussion mostly applies to the simplest of the most common solid oral pharmaceutical dosage forms or "pills", which are conventional tablets and capsules.

More complex controlled release oral drug dosage forms come in a wide variety of formats, and utilize a variety of more highly complex technologies in processing and numerous additional specialized components in their formulation and manufacture. These technologies may be used to achieve different drug release patterns, dosing intervals and release rates for the active drug when a dosage unit is taken. This is important because controlled release drugs are formulated to deliver a safe but effective amount of drug in one dosage unit quantity usually over an 8-, 12- or 24-hour period of time.

This requires the incorporation of multiple doses of the active drug in a tablet or capsule formulated to release a therapeutic dose of the drug immediately upon taking and the balance of the drug in the dosage form at a rate to sustain blood levels sufficient to maintain drug activity over the extended 8-, 12- or 24-hour period of time drug action is designed to be sustained. Consequently, if the intact controlled release dosage unit (tablet or capsule) is broken, crushed, heated or moistened with alcoholic beverage or other beverage with incompatible chemistry, the dosage unit may release an overdose and elicit side-effects which can be harmful, even serious, to the consumer.

Pill Knowledge Is a Wonderful Thing

All too many consumers are all too casual in taking and handling their prescription and non-prescription medication. In health care, the adage "knowledge is power" particularly applies. The following discussions in this section are provided to assist the reader in becoming better informed about topics related to the composition, properties and functions of prescription drugs and dosage forms not frequently discussed with the average consumer. Becoming an educated consumer has many benefits, not the least of which is that it could save one's own life, or that of a loved one.

It will, first of all, enable consumers to communicate better with their health care providers, including their primary care provider, other prescribers, nurses, pharmacists, as well as with family members and associates about personal health matters. This is the foundation for better understanding and communicating how to use medications for best results, as well as for reducing costs for prescriptions written and filled for themselves or members of their family.

To accomplish our objectives, remember that a drug is the active ingredient made available in a dosage form, irrespective of the type dosage form a prescription calls for. Active drug substances are classified according to a variety of criteria and each of these is important to a consumer's understanding and ability to communicate with one's health care providers. Some knowledge of drug classifications also helps consumers maximize the benefits of their investments in these drug products while minimizing the risks associated with taking them.

The subject of pharmaceutical dosage forms will be returned to and discussed later, covering the various types of drug dosage forms, their construction, some of their distinguishing characteristics, their functions, their routes of administration, and limitations, or restrictions associated with the various types. This will help consumers to better appreciate the reasons a pharmacist's admonishments not to crush or break certain tablets or capsules, or to refrain from taking certain products with specific liquids or certain foods, or to store them away from light, heat or cold, or to take them a minimum number of hours before or after eating milk products or taking antacids and the like.

Active Drugs Go to Class

Drugs are classified according to many criteria which are useful in chemically assaying; differentiating between types of drugs based on patent status; determining whether or not a drug is restricted to use under supervision of and by virtue of a prescription order from a licensed doctor, or whether the consumer can independently purchase a drug and self-medicate; or describing drug source or origin, its predominant chemical or physical properties, or its therapeutic or clinical uses; its dosage form type (or drug delivery system); its route of administration; the immediacy of drug release from the dosage form; or its abuse potential. The most important classifications of drugs from a practical drug use viewpoint are covered in the following discussions.

Classification by patent status

The original manufacturers (patent holders) of new drugs promote and sell their products by their *brand names*, which are proprietary and protected by patents and trademarks. Patents protect the active drug substances while trademarks protect brand names. However, *all drugs can be identified by their generic name* whether they are brand-name or generic drug products. Prescription drugs are frequently prescribed by their generic names by some physicians even when they are commercially available only by their brand names. In fact, some brand-name drugs are marketed by different manu-facturers under different brand names. The generic name, however, always remains the same for these different branded products.

After expiration of original drug manufacturers' patents, however, man-ufacturer no longer enjoy a monopoly on marketing the drug product but they do maintain ownership of the brand names for these products. After patent expiration, however, other manufacturers can obtain FDA approval to market their own generic versions of these brand-name products. These generic drug manufacturers, too, can assign brand names to their products but the vast majority merely sell their generic drugs by generic name only. Thereafter, a former single-source, patent-protected new drug is commer-cially available by both its brand name, although it no longer enjoys patent protection, and its generic name.

Classification by level of medical supervision required

The Food and Drug Administration (FDA) is the government agency which evaluates and approves drugs for marketing in this country and which determines whether these drugs will be restricted to sale by prescrip-tion only or whether they can be purchased without a prescription. If it is determined that a drug can be available without a prescription, it is referred to as over-the-counter (OTC) or non-prescription medication. In making its decision as to whether a medication can be sold without prescription, the FDA makes an assessment of whether or not it may safely be taken by patients without the need for medical supervision.

Classification by chemical properties

What different chemical categories or classes of drugs are available? There are many different chemical classification types for drugs or medicines available today. Some classifications are based on their chemical reactivity, acidity or alkalinity, whether carbon-based (organic) or inorganic (not car-bon-based) or based on chemical moieties within broader chemical group-ings. In fact, there may be several different chemical types within a single therapeutic category. For instance, in the therapeutic category of antibiot-ics, chemical types include penicillins (Amoxil or amoxicillin), tetracyclines

(Minocin or minocycline), macrolides (Biaxin or clarithromycin), quino-lones (Levaquin or levofloxacin), etc. This form of classification will not be discussed further because of its highly technical nature and the scope of classification and subclassification involved. For the purposes of this book, chemical class usage will be explained on a case by case basis as it arises.

Classification by physical properties

For consumers, consideration of physical properties of drugs and dosage forms have much more practical application. Physical properties include the physical form of the drug and physical characteristics like form, color, texture, etc. Accordingly both drugs and pharmaceutical dosage forms can be classified as solid, semisolid, liquid or gas. Textures of the surface of solid particles and finished dosage forms can be of several complexions like smooth, granular, slippery, porous, crystalline or amorphous, etc. With respect to their ability to dissolve in water or other most commonly used pharmaceutical solvents, they can also be classified as being soluble, slightly soluble, insoluble, etc.

Classification by therapeutic or clinical use

Of particular importance are the *physiological* or biological properties exhibited by a drug when it is taken by mouth (orally), inhaled, or applied to the skin or mucous membranes, or injected or otherwise administered by or to a person. The beneficial effects from drugs taken internally, injected or applied to the skin or mucous membrane or otherwise, are referred to as *therapeutic, pharmacological or medicinal effects*. Drugs are classified into *therapeutic or pharmacological categories* according to these beneficial effects.

Drugs are officially classified on the basis of their therapeutic actions as approved by the Food and Drug Administration (FDA) when New Drug Applications from the manufacturer are filed and approved. Their approved therapeutic or clinical uses are referred to as approved *indications*.

What are some of the various therapeutic or pharmacological classes of drugs available? Examples of such classes of medicinal drugs based are *analgesic* (pain reliever), *anti-diabetic* or *hypoglycemic* (to lower high blood sugar in diabetes), *antipyretic* (to reduce fevers), *antibiotic* or *anti-infective* (for infections), *antidepressant* (to treat depression), *anti-allergy* or *antihistamine* (for allergies and allergic reactions), *anti-hypertensive* (for lowering high blood pressure), *bronchodilator* (for asthma), *antitussive* (to suppress coughs), and *diuretic* (to remove excess fluids from the body). Simple definitions and examples of these are presented below.

Antibiotics fight or prevent bacterial infections. These may also be referred to as anti-infective agents and are available in a wide variety of chemical classes. Examples of antibiotics are Amoxil (amoxicillin); Biaxin (clarithromycin); Erythrocyn (erythromycin), Levaquin (levofloxacin) and Tetracyn (tetracycline).

Anti-diabetic drugs are used to lower blood sugar in patients with diabetes. Examples are Actos (pioglitazone), Avandia (rosiglitazone), Exubera (the first inhaled insulin powder), Glucophage (metformin), Micronase or Diabeta (glyburide), Prandin (repaglinide), and various forms of insulin.

Analgesics (pain killers), such as Tylenol (acetaminophen), Bayer Aspirin (aspirin or acetylsalicylic acid), and Motrin (ibuprofen), are used to relieve mild to moderate pain.

Both aspirin and acetaminophen reduce fevers and, therefore, are also classified as antipyretics.

Anti-allergy medications are products which relieve allergies by blocking the release of histamine in the body and are, therefore, also called *antihistamines*. These include Allegra (fexofenadine), Benadryl (diphenhydramine), Claritin (loratadine), and Zyrtec (cetirizine).

Theophylline, an asthma medication, relaxes the bronchial tubes and is thus classified as a bronchodilator.

Anti-hypertensive or blood pressure medications reduce high blood pressure. Examples are Benicar (olmesartan medoximil), Norvasc (amlodipine), Tenormin (atenolol) and Dynacirc (isradipine).

Diuretics such as Lasix (furosemide) and Hydrodiuril (hydrochlorothiazide) act by removing excess fluid from the body. In so doing, they also reduce blood pressure and can also be classified as anti-hypertensive drugs.

Listing and defining all of the major therapeutic classes of drugs is beyond the scope of this book. Only a few examples are given to merely illustrate therapeutic classifications of drugs.

All drugs have both beneficial and adverse effects, depending on the dosage taken, the individual patient's sensitivities, existing or prior health conditions, present age, etc. Undesired effects of a drug are referred to as *adverse effects, side-effects* or *toxic effects*. At times, what would normally be considered an unwanted side-effect of a drug may be used by the prescribing doctor to achieve a desired therapeutic outcome in a particular patient. In other cases, a drug may have two or more uses in therapy. An example of this use would include anti-allergy medications (antihistamines) such as diphenhydramine (Benadryl), whose major side-effect is drowsiness, to treat some patients who have problems falling asleep. That is one of the reasons that sometimes a pharmacist may not be able to tell a patient *specifically* what a drug is prescribed for unless the doctor writes the diagnosis or reason on the prescription.

Classification by potential for abuse

Both approved and unapproved drugs with significant abuse potential are classified as narcotic or non-narcotic, and as controlled substances in Schedules I through V, depending on whether they have a recognized legitimate use in standard medical practice in America and on their abuse potential. A brief description of the different federal schedules and examples of drugs in each schedule are discussed below. Individual states may place more restrictive designations on certain substances.

Drugs included in *Schedule I* are drugs or other substances with high abuse potential, no currently accepted medical use in the United States, and/or lack of accepted safe use under medical supervision. Examples of Schedule I substances are heroin, marijuana and mescaline. These drugs are strictly illegal under current federal law.

Schedule II substances are drugs or other substances with high abuse potential and accepted medical use in the United States, but their abuse may lead to severe psychological or physical dependence. Examples of this schedule of drugs are Demerol (meperidine), MS Contin (morphine sulfate), and Oxycontin (oxycodone), and products which have these drugs combined with others, in products such as Percocet (oxycodone and acetaminophen combination). Other products include Adderall (amphetamine and dextroamphetamine), Ritalin (methylphenidate), and Duragesic (fentanyl).

Drugs classified as *Schedule III* are drugs or other substances with potential for abuse less than substances in Schedules I and II, with accepted medical use in treatment in the United States, and the abuse of which may lead to moderate or low physical dependence or high psychological dependence. Examples of this schedule include Lortab and Vicodin (both including hydrocodone and acetaminophen), Vicoprofen (hydrocodone and ibuprofen), codeine, and depressants including pentobarbital and secobarbital.

Schedule IV drugs are those drugs or other substances with low potential for abuse relative to drugs or substances in Schedule III and have accepted medical use in treatment in the United States. Examples of these include Darvocet N-100 (propoxyphene and acetaminophen), Librium (chlordiazepoxide), Talacen (pentazocine and acetaminophen), Valium (diazepam), and Xanax (alprazolam).

Schedule V substances are drugs or other substances with low potential for abuse relative to drugs or substances in schedule IV, accepted medical use in treatment in the United States, and their abuse may lead to limited physical dependence or psychological dependence relative to drugs or substances in schedule IV. Tylenol® with Codeine Elixir is a Schedule V drug in most states.

Due to the varying levels of abuse potential associated with them, refills of scheduled drugs are strictly limited:

1. No refills are allowed on Schedule II drugs at all even if the physician requests them on the written prescription. Drugs in this schedule cannot be telephoned in (except in extremely rare situations) or faxed in and can only be prescribed by licensed physicians properly registered with the U.S. Drug Enforcement Administration (DEA).

2.   Schedule III, IV and V drugs are limited to a maximum of five (5) refills in six (6) months and these prescriptions expire within six months of their original filling. In most states, prescriptions for drugs on these schedules may be written, called in or faxed to the pharmacist.

Classification by source or origin

Active ingredients for medications are obtained from natural sources (microbes, animals or plants, or the earth). Others are synthesized or made from combining chemicals in the laboratory. Some active ingredients may be prepared in the laboratory and then subjected to transformation by a living organism. Such a product is said to be bio-synthetic.

Heart medications digoxin (Lanoxin, for example) and quinine are from natural sources: digoxin from the plant *digitalis purpurea* and quinine from the bark of a *cinchona* tree. Regular insulin is also a natural product obtained from the pancreases of cows (bovine or beef) and pigs (porcine or pork), such as iletin or regular insulin. Most insulin in use today is human insulin made from recombinant DNA utilizing either *e. coli* or *s. cerevisiae* bacteria in the biology laboratory. These insulins are biologicals.

Most medicines available today are synthetic drugs; that is, they are synthesized in the chemical laboratory from two or more other chemicals. They include some of the most popular drugs of the day, like atorvastatin (Lipitor); acetaminophen (Tylenol, etc.), ibuprofen (Motrin), metformin (Glucophage), clarithromycin (Biaxin), furosemide (Lasix), hydrochloro-thiazide (Hydrodiuril), paroxetine (Paxil) and many others like these.

Classification by dosage form type

Most drugs can not be accurately measured or safely handled consistently by the consumer without being contained in a specially designed product form. The product form in which drugs are made available for consumer use is the drug dosage form in which an accurately measured quantity of medication is available for convenient and safe administration, storage and handling according to instructions. A dosage form is the physical form (e.g., capsule, tablet, syrup, spray, powder, for example) by which a medication is made available for use by the consumer. A dosage form is a physical system designed to deliver an accurate, safe dose of a drug to the body consistently and conveniently—whenever it is needed. The choice of dosage form to be used may be a matter of consumer preference, prescriber preference, or may be dictated by age, gender, or physical or mental limitation of the patient.

For instance, it is more appropriate to prepare medications for babies and infants in liquid dosage forms rather than in tablet or capsule form, whereas tablets and capsules would usually be more appropriate for teenagers and adults. However, a liquid pharmaceutical product would likely be more suitable for an adult patient suffering from throat cancer.

Capsules (hard gelatin and soft gelatin), tablets, chewable tablets, creams, lotions, ointments, gels, inhalers, solutions, suspensions, syrups, sprays, suppositories, injections, and wafers are all different types of pharmaceutical dosage forms. Due to the limitless ingenuity of pharmaceutical scientists, there will be a ever expanding number of distinct dosage form types and technologies used to make them.

Dosage forms can further be distinguished by the route of administration (method of application) of the drug to a patient as discussed below.

Classification by route of administration

Drugs are prepared in dosage delivery systems which enable them to be taken by or administered to a patient by alternate routes. Based on this consideration, pharmaceutical dosage forms are also *classified in terms of their routes of administration.* For instance, medications may be classified as:

1. *Oral* products are to be taken internally by mouth and include tablets, capsules, wafers, and various types of liquids such as solutions, elixirs, syrups, and suspensions.

2. *Topical* drugs are to be applied on the skin or mucous membranes, and include creams, gels, lotions, ointments, powders and sprays.

3. *Intravenous* or IV dosage forms are sterile products designed to be injected into a person's veins with a syringe and needle, and provide the fastest onset of drug action.

4. *Intramuscular* or IM products are designed to be injected into the layers of muscle tissue—but are not suitable for injection into the veins—with a syringe and needle.

5. *Intra-nasal* drug products are for application in the nose and include aerosols, nasal sprays, drops and gels.

6. *Intra-vaginal* products are for use in or on the labia of the vagina, like vaginal creams, douches, gels, and suppositories.

7. *Ophthalmic* drugs are for use in the eyes or the eyelids, and include sterile drops and certain ointments.

8. *Otic* dosage forms are drugs to be instilled into the ear canals, to treat pain and infections, and include ear drops.

9. *Rectal* products include enemas, foams, creams, suppositories and ointments are designed to be inserted into or applied to the rectum.

10. *Subcutaneous* (SC) products are suitable for injection under layers of the skin.

11. *Sublingual* (SL) drops or tablets are dissolved under the tongue for certain medications which may be more completely absorbed into the body from this location or may be destroyed in the acid environment of the stomach if swallowed.

The speed at which a drug is absorbed into the body is frequently an important consideration. Further, the degree to which the dose of drug in the dosage form is absorbed and available for action at the site where it is needed is always important. Consequently, drug dosage forms utilize various types of pharmaceutical technology to modify both the immediacy of drug availability and the extent to which the available dose of drug in the dosage unit is released and available for action in the body. In this respect, drug product dosage forms may further be distinguished or classified according to their speed and pattern of drug release.

Classification by dosage form drug release speed or pattern

- *Immediate release* dosage forms are those from which the full dose of drug is released immediately or within a short time period (for instance, 30 minutes) after taking. Most oral pharmaceutical dosage products fall into this classification.

- *Sustained release* (SR) dosage forms provide the immediate release of an effective dose of the drug and then a continuous release of smaller amount of the drug to maintain a therapeutic dose of the drug in the blood over the desired period of 8-, 12- or 24 hours. This is desirable for drugs whose duration of action is short enough to otherwise require more dosing with an immediate release dosage form. Using SR technology in this case reduces the frequency of dosing and also minimizes potential side-effects from the active drug.

- *Long-acting* (LA) dosage forms are a class similar to sustained-release products except that various kinds of drug release patterns may be involved after the initial dose. Technically, sustained-release products (above) and repeat action products (below) are also long-acting dosage forms.

- *Prolonged release* dosage forms are also *long-acting* drug products.

- *Enteric coated* (EC)tablets or capsules are treated with special coating to prevent disintegration of the tablet or capsule or dissolving of the active drug until it passes through the acid environment of the stomach into the more alkaline environment of the small intestines. This prevents irritation of the stomach for certain drugs or, for others, makes possible a more thorough absorption of the drug into the body due to its chemical properties.

- *Film coated* (FC)tablets are simply coated with a thin film of a polymeric material used to make the product easier to swallow, more stable, and to improve appearance of the product. Such a coating may serve dual purposes. For instance, a film coating may also be comprised of enteric coating material, delaying drug release until a tablet or capsule enters the intestines.

- *Repeat action* (RA)tablets usually provide two doses of medication, with the first dose being released immediately and a second dose being released hours later in order to reduce the frequency of dosing for the consumer.

- A virtually unlimited number of other drug dosage formats used to increase patient convenience, compliance with dosing schedules, brand product differentiation, and product protection from environmental influences of moisture, heat, and light.

Dosage form properties such as tablet size, shape, coatings, conventional vs. SR or LA, repeat action, as well as convenience, distinguishability from other drug dosage forms, aesthetic appearance, stability, and patient and doctor acceptance are important characteristics designed and built into the pharmaceutical dosage form. All ingredients used to achieve these properties must meet official compendial standards and be subjected to official testing procedures.

Classification by compendial status

Finally, we come to a most important classification of drugs and drug products with respect to consumer drug safety; that is, a drug's classification as official or non-official by the official compendia of the United States. In order to be considered unadulterated and properly branded in the United States, drugs must conform to standards in the nationally recognized official compendia, which is the result of a public-private collaborative process designed to ensure quality in drugs.

Drugs, dietary supplements, excipients, flavoring agents, colorants, solvents and other inactive ingredients used in the preparation, formulation, manufacture, and packaging of prescription and non-prescription drugs enjoy status of being *official,* if included in one of the legally recognized compendia of the United States. These compendia are the *United States Pharmacopeia* (USP) and the *National Formulary* (NF), now combined into a single volume publication (USP-NF). These compendia are designated as *the official compendia* of the United States in the Federal Food, Drug and Cosmetics Act.

Drug products not recognized in the USP as being official or meeting standards of the USP *are considered to be adulterated and misbranded.* Consequently, counterfeit drugs are adulterated and misbranded drugs. The dangers involved in marketing and taking counterfeit drugs, then, are great because there is no standard of reference with respect to the accurate identity, presence or absence, or therapeutic or toxic quantity of active or extraneous material in the drug products or the packaging in which they are sold. Further, there is no assurance that the labeling on the container is accurate and complete.

Drug Classifications—Why Bother?

Motivated consumers are eager to become drug smart consumers and will want to know as much as possible about how the various drug properties affect their safety and effectiveness, without having to know all of the technical details. They will be eager to understand the importance of a drug's basic properties, indications for which they are prescribed, and major types of dosage forms. In terms of the affordability of drugs, the most important classifications to understand are:

1. Selection of prescription drug product by brand name (sole source) or generic (multiple source product) name represents the most significant avenue for consumers to reduce prescription drug costs. After all, *the patent status of a drug is the major determinant of the price* consumers and other payers are charged for prescription drugs.For generic drugs, competition among various suppliers influence the prices and costs to drug consumers. Read more about this in the chapter on industry profits and consumer interests.

2. Knowledge of therapeutic or pharmacological classification is central to long-term positive drug therapy outcomes, consumer empowerment, and drug safety.

3. Familiarity with drug classifications based on abuse potential is obviously important to patients who take powerful pain medications as well as medications for treatment of sleeplessness (insomnia), anxiety, and depression.

4. Selection of *dosage form type often affects drug costs, patient compliance with drug therapy, and ultimately drug therapy outcomes* and patient safety in variety of ways.

Having some familiarity with the other classifications of prescription and OTC drugs can also be useful in ways that are sometimes obvious by their type of classification.

The main point to remember is that consumers should learn as much as possible about the drugs they take—both prescription and OTC. They can do this by always reading product labels and the drug information provided by the doctor or pharmacist, and by asking the doctor or pharmacist questions about their medication. Other available references include the **Physician's Desk Reference**, which includes the FDA-approved labeling for prescription drug products; the **Merck Manual Home Edition**, which is available in major book stores or direct from the publishers; manufacturer package inserts which are obtainable from product Web sites on the Internet and from some pharmacies; from other independently published online drug information references; or by calling or writing the FDA, or from FDA web sites, for those who own or have access to computers and the Internet. Major pharmacies and many prescription insurance plans also provide drug information on the Internet.

Chapter 3: Eternal Quest for New and Safer Drugs

In order to provide the patient with drugs that are both safe and effective, the U.S. Food and Drug Administration is charged with the responsibility of regulating the pharmaceutical manufacturing industry. It is the pharmaceutical manufacturing firms, however, which discover, research and develop, and market new drugs. Their vehicle for obtaining FDA approval to market new prescription drugs is the New Drug Application. An abbreviated NDA approval process is used for marketing generic substitutes for innovator company, single-source drugs, after expiration of innovator drug patents.

According to FDA estimates, this takes an average of 8.5 years and costs on the average about $500 million from the time formal interactions with a manufacturer begins.[2] However, according to the Pharmaceutical Research and Manufacturers of America (PhRMA), the entire process from the time a manufacturer initiates new drug discovery to FDA approval takes 12 to 15 years and costs around $802 million on average.[3] Following a rigorous review by the FDA, the new drug product may be approved for marketing if it is shown to be both safe and effective in humans. A weighing of risks versus benefits of a new drug figures prominently in the approval decision, particularly for drugs where the potential for adverse effects is unusually great. Also taken into consideration are a company's manufacturing practices, processes, procedures and standards. When an NDA is rejected, millions of dollars have already been invested in the discovery, development and clinical testing of the product.

The hazards involved in new drug discovery, research and development were aptly illustrated by the early 2007 termination by the world's largest drug manufacturer, Pfizer, of advanced Phase III clinical studies of its potential new blockbuster drug, torcetrapib. The clinical trials were stopped after researchers found potentially serious heart-damaging effects from the drug. By this time, the company had already invested some $800 million in developing the new drug product in an unusual combination with its existing blockbuster cholesterol drug, Lipitor, which is scheduled to lose its patent protection soon.

The approval of new drugs by the FDA for legal marketing of these products in the United States can take several different paths and, therefore, affect the costs and total review times for different kinds of new drug products. Still the basic mode for marketing approval is submission of a New Drug Application (NDA) or, in the case of biologics, a Biologic License Application (BLA). NDA submissions are classified by the FDA according to whether the new product is a(an):

- New molecular entity (NME), which is an active drug substance never before approved for marketing in any form in the United States.
- New salt of a previously approved drug.
- New dosage formulation of a previously approved drug.
- New combination of two or more drugs.
- Duplication of an already marketed drug by a new manufacturer.
- New indication or use for an already marketed drug.
- Already marketed drug without a previously approved application.
- Existing prescription drug to be switched to non-prescription drug status.

Then the FDA determines the priority for reviewing an NDA or BLA using two basic criteria:[4]

1. *Drugs which represent significant advances in therapy* over already marketed products *receive priority reviews*, with an FDA goal of

reviewing 90 percent of these applications within six months or less.

2. *Drugs which are similar to currently marketed drugs receive standard reviews*, with an FDA goal of reviewing 90 percent of these within 10 months or less.

Occasionally, a company or other sponsor will submit an application for a new drug to treat or prevent a serious or life-threatening disease. The FDA will accelerate the review and approval of such a drug on the basis of the observation of a significant promising effect, although the manufacturer or other sponsor must conduct additional clinical studies to substantiate the long-term benefit of the drug. This program is referred to as the *Accelerated Approval* program.

Another program to expedite drug approvals is the FDA's *Fast Track Development* process wherein the agency facilitates the development of and speeds its review of a new drug or biologic which demonstrates potential for filling unmet therapy needs for serious or life-threatening conditions.

New Drugs Approved Since 2000

According to the director of the Center for Drug Evaluation and Research (CDER), the FDA approved only seven new molecular entities (NMEs or basic new drug substances) which represented significant improvements during 2005. Further, according to the CDER director's *2006 Report to the Nation*, the median times for approval of priority and standard new drug applications were 6 months and 13.1 months, respectively.

Following is a summary of approvals of new drugs and those representing new medical entities, including those which received both priority and standard reviews, from 2000 through 2006:[5]

- 2000 Out of 98 NDAs approved by FDA, 27 were new medical entities.*

- 2001 Out of 66 NDAs approved by FDA, 24 were new medical entities.

- 2002 Out of 78 NDAs approved by FDA, 17 were new medical entities.

- 2003 Out of 72 NDAs approved by FDA, 21 were new medical entities.

- 2004 Out of 119 NDAs/BLAs approved by FDA, 31 were new medical entities.**

- 2005 Out of 80 NDAs/BLAs approved by FDA, 20 were new medical entities.

- 2006 Out of 101 NDAs/BLAs approved by FDA, 18 were new medical entities.

*A new medical entity is simply a completely new drug which has never been approved for marketing in the United States before in any form. **Beginning in 2004, BLA approvals were combined with NDA approvals.

The Research and Development Process

Physically and conceptually, the discovery of basic new drugs begins in a laboratory, whether that laboratory is a brick-and-mortar structure somewhere at a pharmaceutical company or university research laboratory, or at some remote landscape or ocean depths from which samples of soil or plant or animal life are taken. Sometimes the laboratory is called Serendipity where, by chance, a miracle is sighted, appreciated and grasped before it escapes recognition by the scientist genius. New drug discovery and innovation comprise the backbone of the nation's viable, robust drug industry.

The new drug development process consists of a variety of laboratory research including analytical and synthetic chemical research; physico-chemical research; various biological research including animal pharmacology and toxicology; pharmaceutical dosage formulation research; and clinical research starting with a few but ending with testing in thousands of human patients in three sequential major phases. These extensive and expensive research activities are undertaken to prove that a new drug is both safe and effective prior to its approval for marketing.

The overall course of research and development of a new drug product involves many steps, briefly summarized in the following listing:

1. *Laboratory discovery of a new drug entity* through a lengthy process of synthesizing and screening of literally thousands of research compounds in the chemical or biology laboratories.

2. *Preliminary testing and screening of new research compounds* in the appropriate animal species during preclinical research, followed by long-term animal toxicology studies parallel with ongoing drug R&D.

3. *Filing an Investigational New Drug (IND) application* with the FDA to test the new drug candidates in human subjects during each phase of clinical studies.

4. *Initial Phase I clinical testing and screening of promising new research compounds* in studies with paid healthy human volunteers, using some of the newer statistical methods of dose-finding for subsequent studies of the new drug in humans.

5. *Testing of the most promising candidate(s) in a larger group of human volunteers* who have the disease or condition for which the medication is to be used. This is referred to as Phase II, where preliminary target activity and dose ranges for testing in humans are determined.

6. *Testing of the selected drug formulation in a much larger group of humans* who have the disease condition for which the medication is to be used (Phase III). This will usually consist of thousands of patients in the United States and may include patients abroad, all of whom are tested in accordance with an approved clinical research protocol.

7. *Testing in the chemical (or biological) and pharmaceutical formulation laboratories* simultaneously with the above steps to gain exhaustive details of the chemical and physical properties of each new drug substance.

8. *Preparation of pharmaceutical dosage forms of the new drug entity* for use in clinical studies and to provide research data documenting the compatibility and stability of the new drug substance with fillers, binders and other inert substances which may be required to produce a stable, acceptable dosage form of the new drug substance; compatibility with containers and caps for the product; and the chemical and physical stability of the prime product formulations under different conditions of temperature, humidity, and light source and intensities.

9. *Converting small scale laboratory production experience to actual large-scale production* quantities (referred to as pilot production scale-up) for the new drug pharmaceutical dosage form(s) to determine and solve problems which may surface in such large scale manufacturing.

10. *Establishment of acceptable wording to go on the label* for the pharmaceutical product as well as on inside containers of the product to ensure the proper identification and safe and effective use of the product for its FDA-approved medical indications.

11. *Preparation and filing of a New Drug Application (NDA)* with the FDA for review and approval of the new drug product. Included in this application are the collection and analyses of all relevant research and development experience from physicochemical, animal, dosage formulation, and clinical studies, production trials, quality control specifications, analytical methods development, and process validation.

Chemical and Biological Laboratory Research

Most new drug discoveries begin in chemical or biological laboratories where highly specialized scientists plow their trade. Beginning with initial laboratory work, as many as some 5,000 to 10,000 different compounds may be synthesized or otherwise produced and screened over a period up to approximately five years in order to select a smaller, more manageable number of compounds deemed suitable for preliminary testing in animals.[6]

Chemical and biological research departments of pharmaceutical companies prepare thousands of samples of potential new drugs in the laboratory, utilizing scientific methods derived from years of experience and high technology. Many of these chemicals are finally screened in animals for any desirable effects as well as for any toxic effects which would preclude testing in humans. Alternatively, new drug candidates may originate in the research laboratories of our nation's universities or the research laboratories of the National Institutes of Health.

Preclinical Laboratory Screening in Animals

Promising candidates which meet established screening criteria at the discovery level may then be advanced to preliminary testing in animals for the following purposes:

1. to detect any desired medicinal action in the appropriate animal model(s);

2. to determine any undesirable effects, including effects on organ systems, normal physiological functions, and embryos in appropriate animal models;

3. to determine whether the drugs may be safely tested in humans;

4. to determine, by extrapolation, starting doses of the new medicinal agent to be used for initial testing in humans; and

5. to provide data with which to file an Investigational New Drug Application (IND) with the Food and Drug Administration to be able to test the new drug entity in humans.

These animal or pre-clinical studies may involve up to 250 new drug compounds and take on average up to one and a half years to obtain adequate physiological data on the compounds before they can be tested in human subjects.[7] The purpose of these early studies is to determine how the drugs move through living organisms, meaning how they are absorbed, distributed, and broken down in the body, and how they are excreted or removed from the body. These studies usually continue even after human clinical trials begin in order to find any long-term adverse effects researchers should be looking for in clinical trials. Surviving new drug candidates from these preclinical animal studies are further screened to a much smaller, more manageable group of about five candidates to be used in initial screening in humans.

Pre-market clinical trials in humans are comprised of three distinct, progressive stages: Phase I, Phase II, and Phase III. A fourth phase of clinical research followup begins after the initial drug approval by the FDA and the company's initial launch of the new drug product on the market. This is referred to as Phase IV and is intended to monitor post-market drug safety to uncover any rare or longer-term adverse effects more easily noticeable during extended use of the new drug in much larger populations than the clinical trial groups. This phase of post-market safety surveillance studies has been the subject of considerable discussion recently because of glaring lapses in industry and FDA followup, prompting a two-year examination of the problem by the Institute of Medicine at the behest of Congress. There will be a more detailed discussion of this issue in a later chapter.

Dosage Form Research and Design

Before formulation work can proceed, the basic new agent (NME) must be tested chemically and physically in order to determine its relevant properties, its relative stability and its short-term compatibility with common excipients it can be mixed with in order to design the early clinical supplies for use in Phases I and II. These early studies of the chemical and physical nature of the new drug itself is referred to as preformulation research. This information saves time and other resources and speeds the work of pharmaceutical scientists in screening various dosage formulations of the new drug. Still, even more detailed characterization of the new drug entity's properties usually continues in parallel with later clinical studies.

Preformulation is the term referring to testing of a basic new drug entity in the physical pharmacy laboratories to determine its chemical and physical properties as well as the long-term stability of the drug alone and in combination with commonly used excipients (fillers, diluents, etc.) commonly used to produce suitable pharmaceutical dosage forms. It also involves testing leading drug candidates in potential containers and closures under different environmental conditions of temperature, humidity, and light. These tests enable the research pharmacist to prevent major production-related problems in advance, and makes it easier to troubleshoot production problems once the drug is approved for marketing and the manufacturer begins full-scale commercial production of large scale batches of millions of dosage units of the new drug product at a time.

Preformulation and formulation research must begin in advance of clinical studies in order to provide a suitable drug dosage form to deliver the new research drugs to the people who will be taking the drug in these clinical trials. Even when simple liquid solutions or suspensions are to be used in early clinical work, the expertise of the research pharmacist is required to ensure that the drug is in an acceptable dosage form for administration to test subjects.

Pharmacists and pharmaceutical scientists, with expertise in experimental design and the basics of pharmaceutical compounding, devise, test and evaluate numerous variables which can affect the performance of dosage forms when taken internally. Various formulations are designed and studied for each type of dosage form applicable to medically preferred route of administration based on physical and chemical properties of the drug, pharmacokinetics of the new drug, general patient preferences, dosage safety features, convenience of administration, and aesthetic qualities of the product.

During these studies, acceptable fillers and other inactive ingredients required to make such dosage forms must be determined under varied atmospheric conditions (humidity, temperature, agitation, light, etc.) in order to ensure:

1. the most optimally stable dosage form which results in an appropriate shelf-life for each dosage form the new drug is anticipated to be produced.

2. the timely release of the drug from the dosage form for absorption in the body as required for medical effectiveness;

3. the optimal bioavailability of the drug (the extent to which drug is available in the body following its release from the dosage form)

and its absorption into the body from each drug delivery system to be marketed;

4. the appropriate container and closure (top/cap) for each dosage form of the new drug to be marketed, taking into consideration its inertness to potential reactivity with the enclosed medication, safety from moisture penetration, stated temperature ranges, effects of light and agitation on its physical and chemical stability.

Preceding and parallel with clinical studies, other types of testing related to characterization of chemical and physical properties and replicable testing procedures for the leading candidate(s) usually proceed. Such tests may include, but are not limited to, the following:

1. Characterization of the chemical and physical properties of the new medical chemical or biological compound(s), including such things as chemical identity and purity, physical and chemical properties, and stability. Analytical methods must be developed and refined to enable this to be done accurately and for quality control of raw materials which will be required in the production of new drug dosage formulations used in investigational studies and in the finished dosage forms of the new drug candidate for FDA approval and marketing.

2. Chemical engineers must be able to translate the chemist's or biochemist's small scale laboratory sample synthesis into replicable large scale production processes safely, efficiently and economically, and determine equipment, processes and raw materials, yields, facility, equipment, storage and other logistical requirements. Industry reports that the time required to complete these activities is up to two years.

3. Pharmacists and pharmaceutical scientists design, test, and evaluate numerous formulations for each dosage form under consideration, based on the medically preferred route of administration and physical and chemical properties of the drug, its pharmacokinetic parameters, general patient preferences, dosage safety features,

convenience for administration, and aesthetic qualities of the product.

Clinical Research: Three-Phase Testing in Humans

After emerging from preclinical toxicology studies, the best new drug candidates will be tested in human volunteers in three distinct, progressive phases of clinical trials. These are designated as Phase I, Phase II and Phase III.

For testing new drug compounds in humans, an Investigational New Drug application (IND), which includes research data accumulated during the preclinical studies, must be filed with the FDA. Further, investigative research boards or committees at the sponsoring firm or other organization must review and approve the research protocol to be utilized in the clinical research.

Additional details of each of these clinical research phases are discussed in the following sections.

Phase I Clinical Studies

Phase I represents the first and preliminary testing of a new research compound in humans, usually in a small group of healthy human volunteers. Phase I is designed to initially screen for a desired medicinal action and study the course of the new drug in the body of healthy human beings. People are legally protected against being subjected to drug research without their informed consent.

Clinical investigations in Phase I are typically conducted in small groups of 20 to 100 healthy volunteers over periods of six months to a year. The purpose of these studies is to determine how the drug moves through the human organism: how it is absorbed, distributed, and metabolized in and excreted from the body. Such tests are referred to collectively as *pharmacokinetic studies.*

Healthy volunteers are usually recruited to participate in this research phase for purposes of determining::

1. any desired medicinal action in humans;

2. how the drug is absorbed, distributed, and broken down (metabolized) in and finally excreted from the human body;

3. any major undesirable effects in the human body which could preclude further testing in humans;

4. if it is safe and feasible to test the new drug entity in a larger group of human beings who actually suffer from or are amenable to the disease or condition for which the medication may be used either for prevention or treatment.

Further research on the compounds studied is dependent on the outcome of these studies.

Phase II Clinical Studies

Testing of the new drug in a somewhat larger, yet still relatively small, group of subjects who have the disease or condition which the medication will be used to treat, if successful, constitutes Phase II. These studies are typically well-controlled and closely monitored. They are done to uncover any short-term serious side-effects and risks associated with use of the drug and to document the safety and effectiveness of the drug for specific medical conditions. During Phase II studies, researchers seek to determine a dose range for studying the new drug candidates in 100 to 500 volunteers who have the target disease or condition to be treated.

The research investigators will study the five or so new drug candidates emerging from the earlier Phase I studies to ascertain any clinical activity against the target disease condition. Among the principal goals of Phase II testing are the following:

1. to initially determine the effectiveness of the drug for the particular indication in patients with the disease or condition the drug is intended to treat; and

2. to identify any major short-term side-effects or other adverse reactions in human subjects.

These studies typically span an average of six months to a year, after which the most promising drug candidate is advanced to the third and final phase of clinical studies required before marketing a new drug.

Phase III Clinical Studies

After determining the most promising new drug candidate based on research conducted during Phase II, a company is now able to test its primary new drug entity in a much larger number of patients with the disease or condition for which the drug is intended to be marketed. This is where the best and final new drug candidate is tested on a large scale for its effectiveness and safety in a large and diverse enough group of volunteers to provide adequate data for FDA approval.

Volunteer patients participate in these clinical trials in the United States or abroad under the supervision of approved clinical researchers who specialize in the therapeutic area applicable to the drug being studied. These studies routinely involve up to approximately 5,000 patients and numerous research physicians who must follow the carefully designed clinical research protocol filed with the FDA. From 1,000 to 5,000 volunteers with the target disease condition are usually tested in scientifically designed double-blind, placebo-controlled trials over periods up to about four years. In rare cases, the positive results can be so dramatic that these studies may be cut short to allow all suffering patients in the trials to benefit from the new drug candidate.

Clinical research results are intended to enable the pharmaceutical company or other sponsor to collect sufficient information to document the effectiveness and safety of the drug for use in the treatment of humans with a significant advantage of benefit versus risk. Results from these studies will include:

1. Detailed identification and documentation of major actions of medical significance and major side-effects or adverse effects in humans, including any:

    a. harmful and beneficial effects on individual organs and physiological processes in the human body.

    b. harmful and beneficial interactions with other medications in humans.

    c. harmful and beneficial interactions with certain other disease states.

    d. harmful and beneficial interactions with certain foods or alcohol.

2. More extensive mapping of the course, rate, and extent of absorption, distribution, and metabolism of the drug in the human body, including identification of products of metabolism, and excretion of the drug and any of its principal breakdown products from the body.

3. Determining appropriate dosing for effectiveness and safety to include:

    a. starting and maintenance doses of the drug.

    b. maximum safe and effective dose of the drug.

    c. onset and duration of drug action in the body.

Important considerations in initiating or continuing to develop and market a new drug candidate are:

1. whether the new drug candidate provides any particular clinical advantage over similar drugs already on the market;

2. what side-effects are experienced with the new drug candidate and whether these are fewer or less severe when compared to similar drugs already on the market;

3. whether the benefits of the new drug outweigh the potential risks to patients who will need to take the drug; and

4. whether the costs for the new drug will be justified by the benefits to be derived from its use.

Filing a New Drug Application for Marketing Approval

The mechanism through which the pharmaceutical manufacturer seeks approval to market a new drug in the United States is the New Drug Application (NDA). In order to obtain the approval the manufacturer seeks, a pharmaceutical company files the NDA with the following documentation:

1. pre-clinical or animal test data and analyses;

2. clinical (human) test data and analyses;

3. drug formulation studies, dosage form stability documentation, and other drug information;

4. descriptions of manufacturing procedures, including quality assurance standards and procedures; and

5. facilities and equipment used for all steps and stages in the quality assurance, production, packaging, labeling and storage of the finished new drug product.

Review and Approval of New Drug Applications

Perhaps we've gotten ahead of ourselves; so let us return to the earlier stages of new drug discovery and research and development. Prior to initiating each clinical testing phase, an investigational new drug application or IND must be filed with the FDA by the pharmaceutical company, providing all preclinical, clinical, chemical, pharmaceutical formulation, packaging and labeling, quality control, and other relevant composition, and analytical data available to date.

Based on the clinical effectiveness and adverse effects experienced with the drug candidate by the test population, it will be determined if adequate data exist to prove safety, efficacy and cost-effectiveness to successfully file an NDA with the FDA. This application will typically include compilations of all test data on the emergent new drug candidate into a massive document which have exceeded as many as 100,000 pages or more. These submissions are now made electronically, rather than in hard copy volumes as was done previously.

Such data must prove the physical and chemical stability of the dosage form, the reproducibility and validation of manufacturing and quality control testing procedures and processes, facilities, and equipment, and the effectiveness and relative safety of the new drug in humans:

- in effective dose ranges evaluated for the targeted disease condition;

- in the patient population(s) in which the drug was evaluated; and

- under the conditions patients were tested/treated with the new drug.

If the decision is positive, an NDA is submitted to FDA. According to the PhRMA, the average time for NDA approval for a new drug from time of initial submission is about two years. FDA's count starts from the time that it accepts an NDA filing as being complete for review, which is a shorter period of time.

When an NDA or biological license application (BLA) is filed with the FDA, the agency will conduct either a priority and standard review. To qualify for a priority review, the new drug must represent a significant improvement over existing marketed products for the treatment, diagnosis or prevention of the specific disease or condition for which the drug will be used. A standard review are accorded the NDA or BLA if the new drug or new biologic appears to have therapeutic qualities similar to those of one or more already marketed drugs.

A successful FDA review of this documentation results in approval of the NDA if FDA reviewers' conclude that:

1. the drug is safe and effective for its proposed uses;

2. the benefits of the drug outweigh its risks;

3. the drug's proposed labeling is appropriate; and

4.   the methods used in manufacturing the drug and the controls used to maintain the drug's quality are adequate to preserve the drug's identity, strength, quality and purity.

A generic version of an innovator drug is not legally permissible until the innovator drug manufacturer's patent for the new drug has expired or the brand-name drug manufacturer licenses or contracts with a generic firm to produce a generic product utilizing the innovator company's already approved drug formulation. After drug patent expiration, a generic drug is approved on the basis of an abbreviated NDA (ANDA). Because the effectiveness and safety of the basic new drug entity have been thoroughly documented with clinical data in the original NDA and through medical use through the extended period of patent protection enjoyed by the original producer, the process and expense of gaining FDA approval to market a generic version of a drug are significantly less. These applications are termed "abbreviated" as they do not require the manufacturer to conduct clinical research studies to prove the safety and effectiveness of the drug.

However, the generic drug manufacturer must establish that the generic is chemically and biologically equivalent to the original or pioneer brand-name drug. On approval, the generic manufacturer may market the generic drug product, once patent protection and any extensions of exclusivity for the original manufacturer of the brand-name drug or first ANDA filers have expired.

Market Launch and Post-Market Safety Studies

Finally, that long-awaited moment arrives when the FDA informs the pharmaceutical manufacturer of its approval (or rejection) of its NDA. In anticipation of approval, the manufacturer's pharmaceutical research and development scientists have been working closely with pharmaceutical production in producing large production test quantities of drug in the factory before final commercial batches are produced. These are all tested against the quality control standards submitted to the FDA previously. They must then be released by the quality control department before they can be packaged and distributed strategically to warehouses, where they remain until final FDA approval is received. At that moment, the market launch of the new drug product takes place through distribution of the drug to pharmaceutical wholesalers, pharmacies, hospitals and physicians. And a new drug is now available for prescribers to use in drug therapy for their patients in need of treatment or prevention of an illness or disease.

Heavy advertising and detailing of physicians by pharmaceutical representatives take place. More recently, pharmaceutical manufacturers have begun advertising their products directly to consumers via radio, TV and magazines. Followup clinical studies on the new drug continue as Phase IV for the purpose of better understanding and documenting the drug's safety and effectiveness, especially with respect to potential rare but serious adverse drug effects not yet seen, as well as exploring possible additional uses for the new drug product. Manufacturers are obligated to report all new adverse effects uncovered to the FDA.

Orphan Drugs for Neglected Populations

Pharmaceutical manufacturers require a large enough market for their drugs to justify the heavy investments they must normally make in research and development to prepare new drugs for marketing. Certain diseases are found to occur in relatively small numbers of people, whose lives are as important as all other citizens, but whose numbers would not justify the research and development expense required to bring them to market. Without adequate financial incentive, drug manufacturers shy away from developing products where the market is highly limited. This is where the FDA comes in with its special programs to ensure that consumers with rare disease conditions will have drugs for their treatment available when they need them.

For sponsors of drugs for disease conditions which apply to 200,000 or less people, the FDA provides several types of incentives to facilitate their development. These drugs are referred to as *orphan drugs*. To encourage their research, development and marketing, the FDA grants sponsoring companies seven years of market exclusivity, assists in the design of their research, and provides grants up to $200,000 annually to support orphan drug product development. Sponsoring companies also receive tax credits for their product-related clinical research.

At the end of the FDA-granted period of exclusivity for the initial orphan drug manufacturer, other firms may produce and market the drug. The mechanism for gaining approval for marketing the generic is filing an abbreviated NDA as discussed below.

Generic Drugs Follow Patent Expirations

Generic drugs are FDA-approved versions of previously patented brand-name drugs after the expiration of the drug patents. These drugs, which must contain the same active ingredient(s) and potency as the innovator products, are reviewed using an abbreviated New Drug Application (ANDA) process. The ANDA procedure does not require the extensive clinical studies necessary to gain approval for products containing NMEs.

Competition between the generic drug industry and the major brand-name innovators and the powerful brand defenses presented by Big Pharma are treated in Chapter 7, while Chapter 8 is devoted to discussions of the more economical generic alternative. It discusses the value to consumers of generic drug products, both prescription and non-prescription, and compares generic drug products with their brand-name counterparts on the basis of regulatory requirements and procedures, and their quality, effectiveness, safety and differences in cost.

Some Prescriptions Switch to Non-Prescription Status

Non-prescription drugs can be purchased over the counter without prescriptions and, therefore, are referred to as OTC drugs. To be classified as an OTC drug, a product must be determined to be effective and safe enough for use by consumers in self-care at the specified doses for treatment of non-serious, self-limiting health conditions without medical supervision.

Most non-prescription drugs are marketed pursuant to FDA-approved monographs which don't require NDA approval prior to marketing, if they are manufactured in conformity with the applicable FDA-approved monograph. However, there are two circumstances for which non-prescription or OTC drugs are subject to FDA prior approval for marketing.

If a prescription drug is proposed to be switched to OTC drug status, it must reviewed and approved via an NDA submission. Any other new OTC drug, for which an applicable FDA-approved monograph is non-existent, must use the NDA route to gain approval for marketing. Refer to the chapter on self-care, OTC drugs and CAM, for more detailed information on this class of drugs and the regulatory approval requirements and processes involved in switching older prescription drugs to OTC status.

To briefly recap the process for approval of non-prescription drugs, there are basically three mechanisms available: conforming to an existing FDA-approved drug monograph, without the necessity of filing an NDA; filing an NDA to gain approval for an existing prescription drug to be switched to non-prescription drug status; or filing an NDA to gain approval of a new drug for non-prescription, rather than prescription, status.

Further information on the criteria and procedures required for FDA approval of non-prescription or OTC drugs is given in a separate chapter on self-care, OTC drugs, and alternative and complementary medicine.

Chapter 4: Drug Supply Chain and Consumer Safety

American consumers filled some 3.4 billion prescriptions worth $230.3 billion at pharmacies in the United States during 2005, excluding payments for physician drug orders filled in hospitals and, of course, prescriptions dispensed at Wal-Mart pharmacies. These 3.4 billion prescriptions were filled at community and mail service pharmacies, including independents, traditional chains, supermarkets, and mass merchandisers.

Receipt of prescriptions at local pharmacies or through mail service pharmacies represents the end of a highly efficient and relatively safe chain of drug distribution all the way from the original manufacturer to the ultimate consumer of the product. However, the road from concept to new prescription drug discovery, and from discovery of a new drug entity to the dispensing of the prescription to a patient is a long, arduous, complex and expensive one.

This chapter provides a simplistic representation of the physical distribution pathway for prescription and non-prescription drugs from the original manufacturer to the ultimate consumer. It provides an overview of the major players or links in the nation's pharmaceutical supply chain, the roles each plays in bringing pharmaceutical drug supplies to the consumer, and issues related to safety and costs. Each link plays an indispensable role in getting prescription and non-prescription drug products to consumers virtually anywhere in the country rapidly.

The drug distribution process involves pharmaceutical manufacturers, regulatory agencies, wholesale drug distributors, pharmacy benefit managers (PBMs), prescribers, pharmacists and, of course, the ultimate user of the drug product, who is the patient or drug consumer. Since most insurers use PBMs to manage their drug benefit programs, insurers are subsumed in the discussion of PBMs. While PBMs are not a physical link in the chain of distribution, they are very much involved in all aspects in influencing cost and drug plan product mix by negotiating discounts and rebates on the prices of drugs charged by drug manufacturers to the drug plans they manage.

Along the way, literally thousands of individuals participate at various levels in the complex, time-consuming and costly process leading to the final physical presentation of the healing gem to the eagerly awaiting patient at a local community pharmacy or from a mail service pharmacy or now an Internet pharmacy. The advent and proliferation of Internet pharmacies present special challenges to those responsible for maintaining the integrity and safety of the nation's drug supply, especially because of the convenience it offers online users.

Major Links in the Drug Supply Chain

The system through which prescription and non-prescription drugs are physically transferred from their manufacturer to the patient's retail, hospital, clinic or mail service pharmacy is referred to as the drug supply chain. Although our drug supply chain has been surprisingly safe over the years, in recent years it has become increasingly vulnerable to infiltration with counterfeit or otherwise sub-standard drugs. In fact, counterfeit drugs remain a potent threat to consumer safety worldwide, especially in developing countries in Africa and Asia. The extent to which and how the FDA and state regulatory agencies are addressing this threat is discussed later in this chapter.

The Pharmaceutical Manufacturing Industry

For many prescription drug products, the genesis of a new drug may as readily exist in board room decisions of an innovator, brand-name pharmaceutical manufacturer as in the innovator's laboratory. Why? Simply because the business of the industry mandates that it achieve and maintain a competitive edge in one or more niches of the pharmaceutical industry in order to survive and provide its investors with reasonable returns on their investments. Without reasonable returns on their investments, shareholders will simply sell their stock in the less profitable company and invest in other more profitable companies and industries.

Responsible corporate leadership will constantly be shepherding the corporate research effort into those areas where it can do the most good in serving the public health. However, the realities of life mandate that it must also shepherd the corporate research effort into areas where the financial rewards can be self-sustaining and most profitable to their bosses—the investors or holders of stock in the corporation. Without long-term profitability and financial stability, the corporation withers and dies and can no longer provide its healing miracles to the consumers it has long served in the past.

A simplified representation of the chain of drug supply movement in the United States is shown below:

Table I. FLOW OF DRUGS FROM MANUFACTURER THROUGH THE DRUG SUPPLY CHAIN TO THE CONSUMER			
From the Drug Manufacturer ⇒	**To the Wholesaler** ⇒	**To the Pharmacy** ⇒	**To the Patient**
	Direct to the Community or Mail Order or Internet Pharmacy ⇒		
	Direct to the Hospital Pharmacy or Physician's Office ⇒		
START	*INTERMEDIATE* ⇒		*COMPLETION*
Direction of flow of prescription drug product from the original drug manufacturer to the ultimate consumer. ⇒			

Manufacturers of Different Stripes for Brand, Generic and OTC Drugs

The pharmaceutical manufacturing industry includes the group of companies which manufacture single-source brand-name prescription drugs and biological pharmaceutical products, collectively referred to as Big Pharma and represented by the Pharmaceutical Research and Manufacturers of America (PhRMA); the generic prescription drug manufacturers, represented by the Generic Pharmaceutical Association (GPhA); as well as manufacturers of over-the-counter or non-prescription drugs, represented by the Consumer Healthcare Products Association (CHPA). This industry is under regulation by the Food and Drug Administration (FDA), as legend or prescription drugs must be proven to be effective and safe and are restricted by law to being dispensed by prescription only within the United States. Vitamin-mineral supplements and other nutritional products are not defined nor regulated by the FDA as drugs unless they bear label claims for preventing, treating or curing disease.

Big Pharma is the heart of a robust global network of multinational drug companies involved in the discovery, technological innovation, licensing of new technology, clinical testing, manufacturing, product promotion, marketing and selling of prescription drugs to distributors, hospitals and other health care institutions, physicians and ultimate dispensers. Prescription and non-prescription drugs are regulated by the FDA to guarantee their effectiveness and safety before, during and after they exit the drug supply chain.

The prescription drug industry has given our consumers many basic new drug entities which represented major improvements in drug treatment over existing products, other new drug entities which represented only minor improvements or modifications of existing approved drug products already available, and some new drug therapies which issued in totally new frontiers in drug therapy in preventing the occurrence or spread of deadly disease or in the treatment and cure of previously untreatable or incurable disease conditions.

Members of Big Pharma are both praised and vilified because of the industry's monstrous size and its duality of influences on individual and collective health and wealth domestically and abroad. On the one hand, their prescription drug products are showered with accolades for their miraculous preventive and curing powers which, along with continuing advances in diagnosis and treatment, increasingly expand the quality and length of life for millions.

In the process and, indeed, as a consequence of this, consumers who successfully overcome the adverse effects of their illness become more productive both at home and in the workplace. Employers benefit with decreased absenteeism and increased productivity, collectively boosting the national economy. Everyone benefits as a result.

The Top Brand-Name and Generic Drug Manufacturers

The top drug manufacturers during 2006, based on retail dollar sales volume, were: Pfizer, GlaxoSmithKline, AstraZeneca, Novartis, Teva Pharmaceuticals, Merck, Johnson & Johnson, Sanofi-Aventis, Abbott Labs, and Eli Lilly, in that order, according to data provided by Verispan and reported recently in *Drug Topics* magazine.[8] One of these, Teva Pharmaceuticals, is a generic drug firm. Retail sales for these companies were as follows:

Table II. TOP PHARMACEUTICAL MANUFACTURERS IN 2006 BASED ON TOTAL RETAIL SALES		
Rank	NAME OF DRUG FIRM	RETAIL SALES
1	Pfizer	$ 20,413,717,265
2	GlaxoSmithKline	$ 15,794,165,540
3	AstraZeneca	$ 10,899,867,255
4	Novartis	$ 10,108,785,889
5	Teva Pharmaceuticals	$ 9,936,509,259
6	Merck & Company	$ 8,283,001,583
7	Johnson & Johnson	$ 8,263,048,088
8	Sanofi-Aventis	$ 6,149,228,271
9	Abbott Labs	$ 5,992,405,684
10	Eli Lilly	$ 5,842,618,115
Sources: *Abstracted from* data provided by Verispan published in Judy Chi, Top Pharmaceutical Firms, *Drug Topics*, April 2, 2007.		

However, based on the number of prescriptions dispensed using their drug products, these manufacturing companies came out on top during 2006: Teva Pharmaceuticals, Pfizer, Mylan, Novartis, Watson Pharmaceuticals, GlaxoSmithKline, AstraZeneca, Mallinckrodt Medical, Merck & Company, and Abbott Labs. The prescription volumes earning them this distinction are presented in Table III.[9]

RANK	NAME OF DRUG FIRM	PRESCRIPTIONS FILLED
1	Teva Pharmaceuticals	370,827,222
2	Pfizer	257,653,471
3	Mylan	235,582,276
4	Novartis	222,391,671
5	Watson Pharmaceuticals	207,887,484
6	GlaxoSmithKline	104,542,833
7	AstraZeneca	99,732,796
8	Mallinckrodt Medical	82,903,489
9	Merck & Company	81,640,147
10	Abbott Labs	72,684,001

Table III. TOP PHARMACEUTICAL MANUFACTURERS IN 2006 BASED ON TOTAL PRESCRIPTIONS FILLED

Sources: *Abstracted from* data provided by Verispan published in Judy Chi, Top Pharmaceutical Firms, *Drug Topics*, April 2, 2007.

The innovator segment of the pharmaceutical industry expects that several blockbuster drugs will continue growth and profitability for their companies through 2012 at least. These drugs included the following new single drugs, and one combination product combined by its manufacturer with the basic new medical entity.[10]

1. Acomplia for obesity, NDA filed by Sanofi-Aventis in 2006 and awaiting approval, with previously projected sales of $5.4 billion by 2010;

2. Torcetrapib/Lipitor combo for treatment of high cholesterol, with initially projected sales of $5.0 billion by 2010, before the sponsor was compelled to terminate late term clinical trials early

in 2007 because of serious adverse drug effects from the new drug, torcetrapib;

3. Paliperidone for schizophrenia, in late phase clinical trials by Johnson & Johnson as of Fall 2006, with projected sales of $2.5 billion by 2010;

4. Symbicort approved for asthma, for AstraZeneca in 2006, with projected sales of $2.5 billion by 2012;

5. Galvus for diabetes, NDA filed by Novartis in 2006, with projected sales of $2.1 billion by 2012;

6. Exubera for diabetes, approved for Pfizer in 2006, with projected sales of $1.8 billion by 2010;

7. Rasilez for high blood pressure, NDA filed by Novartis in 2006, with projected sales of 1.4 billion by 2011;

8. Actemra for rheumatoid arthritis, in late clinical studies by Roche as of Fall 2006, with projected sales of $1.4 billion by 2012; and

9. Chantix for smoking cessation, approved for Pfizer in 2006, with projected sales of $1.1 billion by 2010.

However, in recent years, the number of new molecular entities, or totally new drugs, approved in the United States has been decreasing and the major drug manufacturers seem to be looking increasingly to the more profitable biologics market, where costs of treatments are generally much more costly and generic competition is not as yet a meaningful threat as it is with conventional drugs.

Some of the major brand pharmaceutical manufacturers also own their own divisions or subsidiaries to produce generic copies of their own patented products as well as of other companies' drug products no longer under patent protection. In what is presently a highly controversial practice, some major drug producers contract with generic pharmaceutical companies to market "authorized generics" in arrangements deemed by some to be anti-competitive. At the time of this writing, an examination of the issue is being initiated by the Federal Trade Commission to determine if such practices are anti-competitive. Authorized generics are discussed in greater detail later in Chapter 7.

As a natural consequence of the Drug Price Competition and Patent Restoration Act of 1984, commonly referred to as the Hatch-Waxman Act, generic pharmaceutical manufacturers are the guarantors of lower drug prices for consumers. This legislation encourages generic drug manufacturers to challenge the validity of drug patents as a means of boosting price competition in the industry by bringing less expensive versions of brand-name products to market as soon as possible after patents expire or are proven to be invalid. It also restores some of the time brand-name companies lose in not being able to capitalize on their drug patents during the periods of time required to research and develop their new drugs.

There are literally thousands of makers of generic pharmaceuticals in the United States. However, the products of the top 10 manufacturers among the thousands account for the majority of drugs used in filling generic prescriptions in the United States during the year.

The Drug Wholesale Industry: Middlemen

There are thousands of pharmaceutical manufacturers in the United States, including Big Pharma brand-name and generic drug manufacturers. Further, there are thousands of health care institutions and medical facilities where prescription drugs are dispensed. and the thousands of the predominant dispensers of prescriptions to the ultimate consumer in community or retail pharmacies. Increased costs would be prohibitive and fewer and costlier value added services would likely be available to the patient if each prescription drug manufacturer had to deal directly with individual hospitals, clinics, home health care facilities and community pharmacy. The pharmaceutical wholesale distributor solves the problem by serving as the middleman between the manufacturer and the numerous prescription dispensing organizations throughout the country.

The wholesaler is responsible for the safe and efficient distribution of prescription and OTC drugs, health and beauty products and other health related products, as applicable, to the multitude of health care facilities, mass merchandisers, grocery stores, and community pharmacies.

Although there are literally thousands of wholesale distributors throughout the United States, the Big Five drug wholesalers are responsible for the distribution of some 90 percent of prescription and OTC drugs, and other health related products to pharmacies and other dispensing organizations.[11] Together, they provide distribution services to more than 16,700 independent community pharmacies, 26,800 chain pharmacies and mass merchandiser pharmacies, 10,000 hospital pharmacies, 350 mail order pharmacies, 10,000 food store pharmacies, 5,000 long-term care and home health facilities, 61,400 clinics and 1,100 HMOs through their 224 distribution centers strategically located throughout the United States. In some cases, the manufacturer may sell directly to pharmacies or health care facilities.[12]

Drug wholesalers must be licensed by the state, usually by the board of pharmacy in each state. In most states, this includes both resident and out-of-state distributors.

The Big Five and Other Drug Wholesalers

The Big Five full-service wholesalers handle some 90 percent of the distribution of prescription medications but there are also regional wholesalers, smaller (sub-regional and specialty) wholesalers, and secondary wholesalers. The Big Five includes McKesson HBOC, Bergen Brunswig Drug Company, and Cardinal Health. While the Big Five and other major wholesalers usually provide the full range of prescription and OTC drug products and other health related items available, many of the smaller and secondary wholesalers handle only specialized lines of products or, as is the case with secondary distributors, may focus on manufacturer discount sales, purchasing large volumes only of discounted products for resale. Secondary pharmaceutical wholesalers have been referred to as the "ghosts" of the wholesale drug industry, since many cannot be identified and no one knows just how many there are.

The Health Distributor Management Association (HDMA), formerly the National Wholesale Druggists Association (NWDA), represents the major wholesalers and other member distributors. Through this organization and their interactions with other professional organizations and regulatory agencies, they create and exchange knowledge about distribution management and improve standards and business processes.

This is facilitated through regular interactions with some 230 or more manufacturer members (prescription drugs, OTC drugs, health and beauty care, and other health related products) as well as affiliate members including advertising agencies, publications, electronic data processing and automation suppliers, and research companies.

The safest wholesale drug distribution system is one which minimizes the physical shipments of or transactions involving a drug between its shipment from the original pharmaceutical manufacturer and its reaching the ultimate user, the patient.

Most commonly, prescription drug manufacturers sell their products to the Big Five wholesalers. These large wholesalers then sell to the ultimate dispensers (retail chain and independent pharmacies, and other health care facilities) and to smaller wholesale distributors. Complications can arise when large or smaller distributors, especially secondary drug distributors, take advantage of manufacturer short-term sales of individual drugs to reduce their inventories or to meet periodic sales targets.

When this happens, these distributors usually will resell their discounted products quickly to their network of customers, including some larger as well as smaller distributors, in addition to their usual dispensing customers. In this case, a drug product may change hands six or more times before it reaches the dispensing pharmacist in the community, hospital or other health care facility. As the number of physical transfers increases, so does the risk of counterfeit or substandard medications entering our otherwise relatively safe pharmaceutical supply chain.

Protecting Our Drug Supply from Counterfeiters

In order to better protect the nation's drug supply from the introduction and possible retail sale of substandard, ineffective and counterfeit drugs, the federal government has established a prescription "drug pedigree" system for drug distribution.[13] This system requires transactions between manufacturers and wholesale distributors and among distributors to be accompanied by an unbroken paper trail giving the complete history of the physical receipt and distribution of prescription drugs from the original manufacturer all the way down to the dispensing pharmacy.

In most states, the "drug pedigree" program will be implemented under the auspices of the state board of pharmacy based on federal guidelines each state is mandated to follow. The federal guidelines provide for the storage and security of prescription drug products as well as for the handling of returned, damaged and outdated drugs. Some states like Indiana, Florida, and California had already begun implementation of the drug pedigree program at the time of this writing. While the FDA had earlier hoped for wholesalers to have an electronic pedigree capability in operation by 2007, the mandatory use of RFID (radio frequency identification) technology for this program has been deferred to still a later date. However, the FDA gave states until the end of 2006 to implement a prescription drug pedigree program, as its deadlines for the initial implementation have been delayed several times before.

FDA's initial focus in this program is on drugs which have a high potential for counterfeiting because of high demand, higher prices and sales volumes, specialty nature of an expensive prescription drug for a serious or life-threatening disease, supply shortage, prior history of counterfeit or diversion activity, prior history of false pedigrees associated with the product, and status of being in a drug class similar to other drugs with high potential market size and dollar value.

Some of the high profile drugs for which counterfeits or fake imitations have been caught by the FDA over the past few years are Lipitor, Evista, Viagra, and Serostim. The expanding threats from counterfeit drugs and the steps being taken to stem the tide are treated more fully in the Chapter on industrial practices and regulatory agency deficiencies.

Pharmacy Benefit Managers and Prescription Costs

Pharmacy benefit managers (PBMs) oversee the processing of some 70 percent of all prescriptions dispensed in the United States every year; handle about 80 percent of total spending on prescription drugs annually, enjoy more than 90 percent participation by pharmacies in their networks in any given area, and manage drug benefits for about 95 percent of all patients with insurance coverage for drugs.

This discussion of pharmacy benefit managers is included in this chapter on the pharmaceutical supply chain because its major players own or are affiliated with mail service pharmacies, which are integral members of the drug supply chain, and because PBMs interact with and, in some ways, affect the players at every level of the nation's prescription drug supply chain.

While an estimated 50 to 60 PBMs operate in the United States, the largest of these process and pay the vast majority of annual expenditures on prescription drugs, and handle the vast majority of total prescriptions filled and dispensed in the United States. In 2003, the breakdown for these activities was as follows:[14]

1. *Based on total annual expenditures* by PBMs collectively: Merck-Medco (now Medco Health Solutions), 24 percent; Advance PCS, 22 percent; Express Scripts, 14 percent; Caremark Rx, 3 percent; National Prescription Administrators, 2 percent; and other PBMs, 35 percent.

2. *By total number of prescriptions handled* collectively per year: Merck-Medco, 22 percent; Advance PCS, 18 percent; Walgreen Health Initiatives, 13 percent; Express Scripts, 12 percent; First Health Services, 5 percent; and other PBMs, 30 percent.

There are several types of PBMs based on size and ownership. In 2003, the large owners of mail service pharmacies included Medco Health Solutions (formerly Merck-Medco), Express Scripts and Caremark. Small and insurer-owned PBMs included Aetna, Inc., Cigna Corporation, National Medical Health Card System, Inc., Restat, LLC, and Wellpoint Health Networks, Inc.

There were also drug retailer-owned PBMs like Eckerd Health Systems (formerly a subsidiary of Eckerd Corporation); PharmaCare Management Services (a subsidiary of CVS Corporation; Rx America (a subsidiary of Longs Drug Stores); and Walgreens Health Initiatives (a subsidiary of Walgreen Company). CVS has since acquired most of the Eckerd Corporation retail stores as of early 2004. Shortly before this book went to press, CVS was successful in completing a merger with Caremark, after a bout of fierce competition with Express Scripts.

Stand-alone retail chain pharmacies owning their own PBMs are CVS Corporation, Longs Drug Stores, Rite Aid Corporation, Wal-Mart Stores, and Walgreen Company. During March 2007, CVS completed merger transactions with Caremark after fierce competition with Express Scripts.

Pharmacy benefit managers are like invisible "middlemen" involved in the overall processing of retail prescriptions and especially in the determination of what patients and their insurers pay at the local pharmacy for their prescription medication. Although PBMs themselves never touch the actual prescription drug product the consumer picks up at their local pharmacy or receive through the mail, these organizations interact with, influence and/or participate at virtually every level of the pharmaceutical supply chain.

They may or may not own their own retail or mail service pharmacies. Whether they own a mail service or other type of pharmacy, their actions influence the prices and co-pays consumers pay as well as the clinical services received by some 95 percent of consumers who have drug coverage in the nation.

A PBM is a company which administers a drug benefit plan for a plan sponsor, which is usually an insurance company or employer. Plan sponsors include the many health maintenance organizations (HMOs), preferred provider organizations (PPOs), self-insured employers, labor unions, federal, state and local government employers and other entities for which the administration of the prescription drug benefit is separated from the administration of other health or medical benefits.

PBMs utilize two basic dispensing channels: both chain operated and independently owned retail pharmacies, and mail service pharmacies. While they offer a variety of cost-containment programs to plan sponsors, among the more prominent services offered by PBMs are prescription compliance programs and therapeutic interchange programs.

One of the most important functions of a PBM is to manage and contain overall costs for the prescription drug benefit for the plan sponsor and for the patient. In order to accomplish this, a PBM utilizes a number of tools as leverage. The principal tools a PBM uses to manage and contain overall prescription drug costs include prices, drug utilization levels, drug mix and several combination programs involving formularies and other mechanisms.

In order to manage prices, a PBM employs dispensing fees, ingredient costs, rebates from manufacturers, establishment and use of pharmacy networks, maximum allowable costs (MACs) and mail service pharmacies. To manage drug utilization, it employs techniques such as cost sharing, limiting quantities of drugs, drug utilization review, prior authorization, and patient and physician profiling. To manage the drug mix for a particular plan, PBMs employ generic substitution, therapeutic interchange and step therapy (utilizing a "fail first" approach).

Cost sharing involves patients making co-pays and/or deductibles and/or the difference between what an available generic drug and a single-source branded drug costs.

Drug utilization review (DUR) is one of several value added programs a PBM offers. DUR includes electronic clinical reviews in real time at the pharmacy, where programs automatically check for drug-drug interactions, drug-disease state interactions, drug-age alerts, appropriate dosage alerts, underutilization, overutilization and other services.

This involves monitoring the patient's compliance with prescribed therapy and controlling the quantities of drug dispensed during a specific time period and it determines the earliest time that a patient can get a prescription filled or refilled based on the history of prior use of the particular drug. It also involves several safety reviews such as checking for any significant interactions between the drug and any other prescription or non-prescription drugs the patient may be taking (drugs listed in the patient's profile); and between the drug and a patient's disease state, gender or age. It checks whether or not the patient is allergic or has a history of hypersensitivity to the drug or other drugs in the same chemical class as is being prescribed.

Value added programs a PBM offers includes electronic clinical reviews in real time at the pharmacy, where computer programs automatically check for drug-drug interactions, drug-disease state interactions, drug-age alerts, appropriate dosage alerts, underutilization, overutilization and other services.

Prior authorization is applied to selected prescription drugs with limitations due to costs and/or the availability of less expensive drugs which have similar therapeutic effects. A PA-restricted drug requires that the prescriber contact the PBM or insurer with a medical justification for the need for the prescribed drug rather than another non-restricted drug on the formulary.

Drug mix refers to the overall numbers or percentages of the less expensive generic drugs versus the more costly single-source branded drugs available to the consumer through their drug plan. The plan's drug mix helps achieve cost control through balancing the mix of generic and brand-name drugs available through the sponsor's prescription drug plan.

Formularies and disease state management programs are regularly used by PBMs. A formulary is simply a list of drugs which are covered by a particular insurance plan. The variety of formularies available include the *closed formulary*, where access to drugs not listed in the formulary require medical justification before prescriptions can be filled at the plan's co-pays and prices; or the *open preferred formulary*, where incentives are provided to encourage the use of drugs on a so-called "preferred" list in order for a patient's prescriptions to be filled at the plan's co-pays or prices. In other cases, a formulary may not be used but prior authorization is required for a select list of drugs. In still other cases, neither a formulary nor prior authorization may be required but rather retrospective analysis of physician prescribing may be used. This is referred to as physician-focused retrospective reviews.

Disease state management is a group of programs which target specific disease states or health conditions for which medical costs are abnormally high and, accordingly, provide cost-effective interventions by nurses and pharmacists to help the patient better manage their health through patient counseling, education and other lower-cost activities. Such high profile and costly disease states, such as asthma, diabetes, high blood pressure and high cholesterol are among the most common disease states targeted for these type interventions.

Collectively, the programs offered by PBMs are referred to as managed care. Managed care services usually involve two types of drug substitutions. These are generic substitution and therapeutic substitution.

Generic substitution, which is largely mandated by state law, is subject to expressly indicated overrides on the face of prescriptions by prescribing physicians, and subject to federal law for federally supported programs like Medicaid and Medicare. Generic substitution involves the automatic substitution of an available, FDA-approved, multi-source generic drug when the patient presents a prescription for a brand-name drug to the pharmacist.

Therapeutic substitution, on the other hand, involves the substitution of a "preferred" single-source brand-name drug used for the same disease or condition as the single-source brand-name prescription drug the doctor prescribed for the patient. Such substitutions usually are not automatic but require consultation with and prior approval of the prescribing physician.

PBMs are in a unique position to negotiate significant price discounts and rebates from single-source prescription drug manufacturers in order for their drugs to be placed on drug plans' formularies. This enables them to massage the prescription drug mix of single-source brand-name prescription drugs with multi-source generic substitutes. Consequently, the use of manufacturer discounts and rebates and therapeutic interchange programs have emerged as major issues regarding both drug plan costs and patient costs.

The preferential pricing granted mail order pharmacies, particularly for chronic disease medications, and self-dealing through PBM ownership of the major mail order firms have raised heated questions about these drug benefit managers because, among other things:[15]

1. The top five pharmacy benefit managers each own mail order pharmacies and they control some 80 percent of the mail service prescription processing business;

2. Mail order dispensing is more profitable for them than their income from administration of the drug benefit plans;

3. They receive larger rebates from drug companies on the more expensive single-source brand-name drugs than they receive on multi-source generic drugs;

4. Routine turnaround times for mail order prescription processing allows them to contact and obtain the permission of prescribing doctors to switch prescriptions to alternative single-source brands;

5. Researchers have documented that "switching" or therapeutic substitution occurs more often in mail order pharmacies than in stand-alone or unaffiliated mail service pharmacies; and

6. They can also profit from repackaging prescription drugs purchased in large volume at lower unit costs and pricing these repackaged drugs at higher unit AWPs than the manufacturer's original unit drug charges.

Researchers at the University of Minnesota's PRIME Institute surveyed pharmacies in 10 different major metropolitan areas on the same mix of drugs drawn from the 10 most frequently dispensed therapeutic categories of drugs over a one year period.[16] They compared the percent of all prescriptions filled with generics by payment source. While other factors are also at play, they concluded that disparities in generic substitution rates could not be explained simply by differences in product mix of prescriptions dispensed at regular pharmacies vs. those at mail order pharmacies.

Their results showed the generic substitution rates to be highest for cash payers and lowest for PBM-affiliated mail service pharmacies. Generic substitutions were made for cash paying patients 53 percent of the time; for Medicaid patients, 51 percent; for third party payers, 42.2 percent; for unaffiliated mail order pharmacies, 38.9 percent; and for PBM-affiliated mail service pharmacies, only 29.4 percent.

The Retail Pharmacy Industry: Drug Dispensers

The final stop for the "pill" prior to delivery to the ultimate consumer or care-giver is the hands of the pharmacist at a traditional pharmacy or drug store. It arrives at the pharmacy daily or several times weekly either directly from a wholesale distributor, a regional distribution center of a large chain store operation or directly from the manufacturer in a supply chain designed to protect the integrity of the prescription drug from the time it leaves the manufacturer's warehouse to the time it reaches the hands of the consumer.

During 2005, the most recent year for which these data were available at the time of this writing, patients filled some 3,380,000,000 prescriptions through retail pharmacy outlets throughout the United States. This represented an increase of 3.3 percent in the number of prescriptions filled during 2005 over the previous year. These were filled at some 17,917 independently owned community pharmacies; 21,349 traditional chain pharmacies; 7,146 pharmacies in mass merchandiser stores; and 9,771 supermarket pharmacies. The total number of chain outlets through which these patients were able to fill their prescriptions during the year was 38,266, according to the National Association of Chain Drug Stores (NACDS). Almost 71 percent of these prescriptions were filled at chain store pharmacies (traditional chains, mass merchandisers and supermarkets), while the balance was filled at independently owned and mail service pharmacies.[17]

Prescription Drug Costs for Consumers

Total costs for prescription drugs in the United States are estimated to comprise about 10 percent of total health care costs, according to the Centers for Medicare and Medicaid Services (CMS). CMS includes in its retail drug spending costs prescription transactions from chain and independent community pharmacies, grocery store pharmacies, mail order pharmacies, and mass merchandiser pharmacies.[18]

The overall average consumer cost for filling a prescription in the United States during 2005 was $64.86 compared to $63.19 in 2004 for an increase of 7 percent. The average price the patient or the patient's prescription drug plan paid for a *brand-name medication* in 2005 was $101.71 compared to $95.01 in 2004. The average price paid for *generic prescriptions* rose 3.3 percent from $28.87 the previous year to $29.82 in 2005.[19]

Out of the estimated average prescription cost of $64.96, the pharmaceutical manufacturer received $49.69 or 76.6 percent; the drug wholesaler, $2.03 or 3.1 percent; and the dispensing pharmacy, $13.14 or 20.3 percent. These estimates are based on the gross prescription income distribution for 2004. Most patients, however, pick up their prescriptions for a co-payment which is a fraction of the total price negotiated and paid by the patient's insurer. In most cases, the overall determination of a consumer's costs for a prescription is a much more highly complex matter. It involves pharmacy benefit managers rebates, negotiated dispensing fees, and other such modifications, made seemingly simple at the pharmacy via real time computer interlinks with insurers and/or their PBMs, which instantaneously feed back patient prices and co-payments to be paid for the prescription.[20]

Receiving, interpreting, verifying and monitoring drug interactions, allergies and age, gender and disease alerts for these patients were some 136,773 community pharmacists, of which about 120,000 served through chain pharmacies, about 99,000 of which worked in full-time capacities. All of these pharmacists are licensed by their state boards of pharmacy as registered pharmacists. Many of these pharmacists completed six years of intense university level studies and training in pharmacy, pharmaceutical chemistry, pharmacology and therapeutics, clinical pharmacy and hold the doctor of pharmacy (Pharm.D.) degree. Indeed, they are the undisputed experts in dosage formulation, drug information, drug therapy management and the biology, chemistry and pharmacology of drug products.

People objecting to the price of prescription medication often ignore the costs of the pharmacy in actually handling the various tasks associated with receiving, validating, drug utilization review, patient therapy monitoring, insurance problem handling, etc. Many feel that the pharmacy should charge close to what it has been charged to purchase the drugs required to fill their prescription. However, as in any business, the retail pharmacy must cover its overhead costs as well as make a reasonable profit in order to stay in business to serve its customers.

Because the government finances the largest programs for providing prescription medication to Medicaid and Medicare beneficiaries, its focus on cost savings has resulted in a proposal to reduce pharmacy reimbursement for dispensing costs. Because of this and to provide reliable information to the decision-makers in government and as guidance to the community pharmacy industry, the accounting firm, Grant Thornton LLP, was commissioned by the Coalition of Community Pharmacy Action (COPA). The study used data from some 23,000 community pharmacies and some 832 million Medicaid prescriptions from all 50 states, the District of Columbia and Puerto Rico, in determining the national cost of dispensing as well as state-level costs of dispensing for 46 states. The results of this study showed that the national average cost of dispensing is $10.50 per prescription.[21]

Since dispensing costs vary from pharmacy to pharmacy, from state to state, and according to store dispensing volume, this naturally causes prescription prices to vary from pharmacy to pharmacy and from state to state. However, when all prescriptions filled were given equal weight in the study, the calculated *average cost of dispensing* was $12.10 per prescription. The range of average costs by state varied from $8.50 per prescription in Rhode Island to a high of $13.08 prescription in California. Costs of dispensing do not include the cost of the drug product being dispensed or profit margin for the pharmacy.

The Nation's Top Pharmacies

Of the some 190 pharmacy chains holding membership in the National Association of Chain Drug Stores in 2005, the following 10 drug chains came out on top based on their total prescription dollar sales (See Table IV):

Rank	Company Name	Company Headquarters	2005 Rx Sales	No. Rx Stores
\multicolumn{5}{c}{**Table IV: TOP TEN PHARMACY CHAINS RANKED BY 2005 PRESCRIPTION SALES IN MILLIONS OF DOLLARS**}
1	CVS Corporation*	Woonsocket, R.I.	$ 28,900	6,071
2	Walgreens	Deerfield, Ill.	$ 27,350	4,953
3	Wal-Mart	Bentonville, Ark.	$ 11,036	3,289
4	Rite Aid	Camp Hill, Pa.	$ 10,900	3,323
5	Brooks-Eckerd	Warwick, R.I.	$ 5,500	1,853
6	Kroger	Cincinnati, Ohio	$ 5,450	1,913
7	SuperValu**	Eden Prairie, Minn.	$ 3,600	956
8	Safeway	Pleasanton, Calif.	$ 3,100	1,332
9	Medicine Shoppe	St. Louis, Mo.	$ 2,520	1,066
10	Sears Holdings	Hoffman Estates, Ill.	$ 2,400	1,171

Abstracted from *Drug Store News*, National Association of Chain Drug Stores, Alexander, VA, August 28, 2006, p. 32. *Includes data from 701 stand-alone Osco and Sav-On drug stores acquired from Albertsons. **Includes data for 1,124 stores acquired from Albertsons.

Consolidation within the largest drug store chains continues, increasing the number of prescriptions filled by each enlarged firm, prescription sales volumes, and the number and geographic distribution of the major pharmacy chain units as they absorb the smaller chain pharmacies.

The prescription dollar sales and prescription volumes for the years 1995, 2000 and 2005, are broken down in Table V by type of store housing the pharmacies. Total prescription dollar sales for all types of stores increased from $72.247 billion in 1995 to $230.256 billion in 2005. At the same time, the number of prescriptions filled by all of the store types increased from 2.125 billion in 1995 to 3.383 billion in 2005.

Table V. PHARMACY SALES IN MILLIONS OF DOLLARS AND MILLIONS OF PRESCRIPTIONS FILLED BY PHARMACY TYPE DURING 1995, 2000 AND 2005						
Types of Pharmacies	1995 Prescriptions		2000 Prescriptions		2005 Prescriptions	
	Sales	Number	Sales	Number	Sales	Number
Drug Stores	$ 49,847	1,580	$ 92,537	2,033	$ 136,272	2,305
Chains	$ 28,794	914	$ 61,182	1,344	$ 94,453	1,562
Independents	$ 21,053	666	$ 31,354	689	$ 41,819	743
Mass Merchants	$ 7,663	238	$ 13,542	293	$ 22,480	365
Supermarkets	$ 7,356	221	$ 17,362	394	$ 27,576	469
Mail Order	$ 7,382	86	$ 22,129	146	$ 43,929	244
TOTALS	$ 72,247	2,125	$ 145,570	2,865	$ 230,256	3,383

Abstracted from Commerce Department, IMS Health and NACDS data via Table 30, Pharmacy Sales and Prescriptions, The Chain Pharmacy Industry Profile 2006, National Association of Chain Drug Stores.

Anatomy of a Prescription

Because drugs carry risks of bodily injury and even death when improperly used, those drugs deemed to require medical supervision for their safe and effective use in the general population, are available in the United States by prescription only. As a result, the legal means by which patients can receive prescription medications in the United States is the *prescription,* which is usually a handwritten, verbal, typed, faxed, or electronic communication from the prescribing physician to the pharmacist.

This prescription is an order to the pharmacist for drugs or medicines or mixtures or combinations of these, written, signed or otherwise legally communicated to the pharmacist by a physician, dentist, podiatrist, veterinarian or other medical practitioner licensed to practice in a state and authorized to write prescriptions.

This includes orders for drugs or medicines transmitted through word of mouth, telephone, or other means of communication by a licensed physician, dentist, podiatrist, veterinarian or other qualified medical practitioner. The licensed prescriber must authorize the prescription within the scope of his/her prescribing authority in applicable statutes and/or regulations. For instance, a veterinarian can lawfully write prescriptions for the treatment of animals but is not allowed to prescribe medications for a human patient. A dentist cannot legally prescribe contraceptives according to most state laws.

In order to reduce medication errors, the Institute of Medicine of the National Academy of Sciences and major professional medical and pharmaceutical groups are promoting the expanded use of the electronic prescription. Electronic prescriptions are entered into a computer at the prescriber's office and transmitted directly to the pharmacy's computer, minimizing potential errors of interpretation of poorly scrawled paper prescriptions. In fact, the two most widely available systems for use in transmitting electronic prescriptions merged in 2007 to facilitate uniformity and universality in this process.

Notwithstanding this, prescriptions for Schedule II drugs can be dispensed *only on a written prescription* and *cannot be refilled* even once, even if the prescriber indicates a number of refills on the original prescription. At the time of writing this chapter, prescriptions for Schedule II drugs could not be phoned in or faxed to the pharmacist by the prescriber, except under the most extreme circumstances, in which case prompt followup by both the prescriber and pharmacist must be done, and special unusually stringent handling and record-keeping requirements must be met.

In order to insure that patients have access to the most economical medications available and to contain the prescription cost component of health budgets, most states require generic substitution when a brand-name drug is prescribed. When necessary, the prescriber can indicate that the brand-name medication is required, in accordance with state law. Most insurers also require generic substitution except where a generic is not yet available. If a generic substitute is available, the patient may obtain the brand-name drug, if he or she is willing to pay either a larger co-pay for the drug, the difference between the costs for the generic substitute and the brand-name drug, or the full price for the brand-name drug.

How to Obtain Prescription Medication

In the community, it is the patient's responsibility to consult their health care provider to obtain prescriptions or refills when needed. It is not the responsibility of the pharmacist to do so. Most pharmacists will assist the patient by contacting the patient's physician or other prescriber when necessary and for certain patients requiring assistance due to certain disabilities. Many consumers use the pharmacist to avoid contacting their provider directly themselves and take advantage of the generosity of the pharmacist in this regard. Pharmacists eagerly contact a patient's prescriber on their own initiative when there is a concern regarding proper dosing for the patient, compliance issues, potentially serious drug-drug interactions or patient disease states where the drug should not be used as well as affordability issues.

Outside the hospital environment, in order to have a new prescription filled, a prescriber can call in the new prescription to the patient's pharmacy of choice, fax or electronically transmit the prescription to the patient's pharmacy, or the patient must present the original signed prescription, not a copy or a prescription with a rubber stamped signature on it, to the pharmacist. Patients cannot call in their own new prescriptions to the pharmacy as the pharmacist is legally responsible for determining the legitimacy of each new prescription as well as interpreting the prescriber's instructions in order to fill each prescription accurately for each patient.

Forged prescriptions are illegal and unacceptable for filling. Forged prescriptions are prescriptions written or otherwise communicated to a pharmacist in the name of a legally authorized prescriber by someone other than the prescriber or the prescriber's legally authorized agent.

Why Patients Get Both Generics and Brands

Roughly half of the drugs on the U.S. market have generic equivalents available. All states have laws which permit or require pharmacists to substitute less expensive generic drugs for the more expensive brands prescribed, under the following conditions:

- A generic substitute is available for the brand; and
- The prescriber does not require the brand *and* specifically indicates that on the patient's prescription as stipulated by state law.

This protects the consumer from being required to purchase expensive brand-name prescription drugs when more economical generic alternatives are available. In cases where a generic is available and the physician directs that only the brand be dispensed, the pharmacist can intercede with the prescriber on the patient's behalf when he knows of an acceptable generic or suitable therapeutic alternative more reasonably priced for the patient. If the prescriber approves, the substitution can be made. Otherwise, the prescription must be filled for the brand as originally directed by the physician that it is to be "dispensed as written (DAW)" or "do not substitute (DNS)" in accordance with the applicable state's legal requirements. In this case, a generic cannot be substituted.

To Get Brand-Name Prescription Drugs Only

Consequently, when the patient presents a prescription of a brand-name drug without the designations "DAW", "Dispense as Written", "DNS", "Do Not Substitute" or other similar language, in accordance with the laws of the state where a prescription is being filled, the generic substitute is automatically dispensed, if one is available. In some states, like Florida, these designations must be handwritten by the prescriber or phoned in with such instructions. In such cases, checking a box by the printed designations on the prescription blank or simply signing on the line above printed designations is legally void. In other cases, prescription insurance coverage dictates the terms determining the dispensing of available generics in place of more costly brand-name medication.

When patients wish to purchase only brand-name prescription drugs, they should ask their physician to designate in writing on the prescription that only the brand is to be dispensed. If generics are available for any of the brands prescribed, however, and patients insist on brands, these are some of the consequences:

- Cash-paying customers will pay a higher price for their prescription medication.

- Consumers who have prescription drug insurance will usually pay a higher co-payment and, in addition, will likely pay the difference between what the generic would cost the plan and the higher cost to the plan for the brand-name drug.

- Consumers will get exactly the product they want.

- There will usually be no advantage in therapeutic benefit in taking a brand-name drug over taking the generic product instead.

Unlimited Prescription Refills Merely a Dream

Authorization to refill regular prescriptions cannot exceed one year, irrespective of the number of refills the prescriber writes on the prescription blank or authorizes by phone. For controlled substances in Schedules III-V (alprazalam (Xanax), Klonopin (clonazepam), Restoril (temazepam), products with codeine and derivatives of codeine such as hydrocodone), prescriptions can be refilled for a maximum of five times in a six-month period, irrespective of the number of refills authorized by a physician.

Adding One's Touch to a Prescription Is Illegal

Since a prescription is a restrictive legal document, it is illegal for patients to alter it in any way except to correct the spelling of a name or to correct or add their birth date, address or telephone number. Nothing else, including the patient, the name of the drug, the instructions, the quantity, the number of refills or name of the prescriber, can be changed by the patient or their representatives. To make any such changes on the prescription is illegal and, if done on a prescription for a narcotic drug or other controlled medication, the prescription is void and the patient may be reported to the prescriber and/or to legal authorities, depending on the nature and quantity of the medication being sought and whether or not the patient has a history of abuse or other prior problems in handling similar prescriptions at pharmacies.

Pharmacists Do Much More Than Count and Pour

The pharmacist is the health care professional to whom has been entrusted the duty and responsibilities of dispensing prescription medication which has been properly prescribed by a duly licensed medical practitioner. Because the role of the pharmacist is so vital to the health and safety of our patients, the pharmacist is required to complete a lengthy five or six-year course of professional and scientific studies at an accredited college or school of pharmacy before he/she can be licensed as a pharmacist.

The pharmacist is charged with receiving, analyzing, filling and dispensing the right drug to the right patient in the right dose and with the right instructions for taking the medication. The pharmacist also counsels the patient on proper drug use and provides drug information to the patient. Most of the many functions and tasks the pharmacist must perform, with or without the aid of pharmacy technicians, are virtually invisible to most patients. This is because this health care professional is so highly trained and experienced that s/he is able to perform a long list of tasks related to filling prescriptions efficiently and rapidly.

Patients need to understand that pharmacists cannot just count and pour their medication instantaneously after their prescription is presented to be filled or refilled. Although assistants like pharmacy technicians may perform some of the pharmacist's tasks under his/her direct guidance, the pharmacist still must check, verify and document the accuracy and completeness of the prescription before it is released to a patient. The pharmacist must also be prepared to discuss with the patient any warnings or potentially harmful effects of a medication, especially potentially harmful drug interactions, as well as answer any questions the patient may have at the time of dispensing the medication.

As discussed in a later chapter, distractions are a major cause for errors by prescribers when writing prescriptions and for pharmacists when filling prescriptions. For a better understanding of the many tasks involved in the processing of every prescription the pharmacist receives, the following is provided as an illustration of these. For every new prescription the pharmacist will:

1. Verify that the prescription is legitimate and complete;

2. Obtain and verify relevant personal information on the patient, including age, contact information, address, drug allergies, health conditions and other drugs being taken if this information is not already available to the pharmacist;

3. Interpret the name, dosage form, strength and amount of the drug prescribed to determine:

    a. If the drug (active ingredient) is appropriate for the patient based on age, sex and other criteria and, if not, consult with the prescriber with concerns and recommendations;

    b. If the *dosage form* is appropriate for the patient (e.g., capsule vs. a syrup vs. a chewable tablet for a 4-year old, for instance) and, if not, consult with the prescriber with concerns and recommendations;

4. Interpret the instructions from the prescriber to the patient and evaluate the appropriateness of the prescribed dose based on standard approved product doses;

5. Inform the patient and/or prescriber of any known unusual circumstances which may adversely affect the patient's drug therapy;

6. Create or update the patient's profile with relevant information from the patient;

7. Using the information in the patient's profile, monitor whether the patient is over-using or under-using prescription medication and, if so, consult appropriately with the patient and/or prescriber on the issue; and

8. Evaluate on-file patient allergy information against prescribed medications to prevent dispensing a drug to which the patient may be allergic.

Consumers who casually take prescription or non-prescription drugs as if it were merely M&Ms, should think again and think hard! While this subject will be returned to later in this book, it is important enough to proclaim now before going further that all drug consumers should:

1. Always read the drug label thoroughly.

2. Always follow printed instructions on the label and verbal instructions from the doctor and pharmacist.

3. Always ask questions of the doctor or pharmacist to clarify unclear information about properly using the prescribed medication.

4. Always give that innocent-looking "pill" the respect that it deserves to prevent its wrath of undesirable side-effects and other medication-related problems.

5. Always remember that prescription and over-the-counter drugs are potent chemicals with safety risks.

6. Always take and store prescription and over-the-counter drugs responsibly and safely, particularly around children, as instructed.

Prescription Drug Assistance Is Available

Consumers rarely pay full price for their prescriptions because prescription drug insurance coverage is usually available to reduce consumer costs to co-pays. Co-pays are only a fraction of the total cost for the drugs they receive. Prescription drug insurance may also reduce consumers' out-of-pocket costs for a prescription to co-insurance, which represents a percentage of the total cost of the drugs they receive or, sometimes, a combination of the two approaches. These amounts are usually fixed payments for a month's supply or three-month or 90-day supply of a prescription medication. The amounts paid are usually based on the "tier" classification of a drug on a restricted list of drugs adopted for a particular insurance plan.

The leadership of the trade organization for the nation's brand-name pharmaceutical manufacturers pointed out recently that generic drugs usually make up the first tier of a drug formulary because they are the least expensive.[22]

The listing of drugs and the conditions under which they will be covered by the insurer or other third party is referred to as the *drug formulary*. Some less expensive brand drugs may also be included in a first tier, depending on the prescription drug plan. The more expensive brand-name drugs are usually covered as second and third tier drugs, with increasingly larger co-pays or co-insurance requirements, depending on a variety of factors.

Hospitals, other health care institutions, insurance companies and state health insurers use restricted lists of drugs they will cover, as a means of containing costs in their health care programs. The insurer's formulary may be a single listing of drugs which are covered under the same conditions (for instance, generic substitutes only) or a two-level listing (for instance, generic substitutes require the lowest co-pay, while brand-name medications require a higher co-pay) or a three-level listing, where a third level listing of drugs, such as the most expensive brand-name and specialty drugs, are available at the highest co-pay or full discounted price.

Unfortunately, the uninsured, the under-insured, and Medicare Part D prescription plan recipients who have reached the "doughnut hole," where their plans will pay no more until a catastrophic limit of full payments has been reached by these patients, all fare worse.

While this book is not intended to cover such programs, consumers should know that there are a variety of prescription drug assistance programs (PDAPs) offered by some state governments and major pharmaceutical manufacturers for patients unable to pay for their prescriptions. In fact, according to the national trade organization for these companies, pharmaceutical companies provided millions of dollars in free drugs through their PDAPs in 2005. Through the web site for the Partnership for Prescription Assistance, consumers can access more than 475 public and private patient assistance programs, including more than 180 programs offered by the major pharmaceutical companies. To contact them online, go to *www. pparx.com* or call toll free 1-800-477-2669.

Following is a short list of sources of prescription drug assistance or information on reliable sources as of January 2007:

1. Physicians, pharmacists and local, county and state health departments.

2. The consumer's federal protection agency, Federal Trade Commission, Bureau of Consumer Protection, Office of Consumer and Business Education, Washington, D.C. For free information, consumers may visit www.ftc.gov or call toll free at 1-877-FTC-HELP (1-877-382-4357) or for TTY users, 1-866-653-4361.

3. The Federal Centers for Medicare and Medicaid Services, where Medicare information can be accessed at www.medicare.gov by calling toll free 1-800-MEDICARE.

4. Free Medicine Foundation's help desk is available toll free at 1-573-996-3333 or online at www.freemedicinefoundation.com.

SECTION II: Drug Safety and Related Issues

Chapter 5: Medication Errors and Drug-Related Problems

Drugs are discovered, marketed, and prescribed for their intended beneficial effects in health care of patients and self-medicating consumers. Unfortunately, these wonderful benefits from drug therapy are accompanied by some unwelcome baggage in the form of potential adverse drug reactions. Today's miraculous wonder drugs constantly present challenges to physicians and patients in the form of mild to severe, sometimes fatal, undesired effects when used and all too frequently blighted by preventable man-made errors and, to a considerable degree, far too many self-induced medication-related problems. Any casual use escalates the potential for adverse outcomes from prescription drug use. However, this also applies when non-prescription drugs are involved.

Unlike most common commodities, taking prescription or OTC drugs is not a "one size fits all" proposition by any means. Each individual's genetics and bodily makeup are unique and subject to significant variation in response to drugs. While remarkably safe when prescribed and used according to standard practice guidelines and FDA-approved drug labeling, prescription and non-prescription drugs have the potential for harm to the human body. Because of this, considerable effort and expense are invested in determining and informing consumers and prescribers of potential safety risks associated with taking prescription and non-prescription drugs prior to and following initial FDA approval and marketing of these drugs.

Discussions in this and the following chapters are designed to provide a simple foundation for consumers to better understand how drugs work in the body; the kinds of potential problems associated with the use of prescription and non-prescription drugs; how and why such problems can adversely impact their health and overall health care costs; and how consumers can empower themselves to minimize these risks.

Major Types of Medication-Related Problems

A report of a two-year study by the U.S. Centers for Disease Control and Prevention (CDC) recently disclosed that *adverse drug reactions* cause some 700,000 emergency visits to hospitals annually. Further, the results showed that one of every six emergency room visits resulted in the patient's hospitalization in an inpatient unit or emergency department observation bed.[23]

A 2006 report of the Institute of Medicine focused on preventing *medication errors*. It disclosed that medication errors kill, sicken, or injure at least 1.5 million people in U.S. hospitals yearly and conservatively estimated that treating injuries caused by medication errors in hospitals alone costs $3.5 billion annually.

According to the CDC, it has also been established that virtually all important bacterial infections in the United States and worldwide have become resistant to the most effective antibiotics available for their control. Some 2 million patients get infections while in a hospital, of which some 90,000 reportedly die annually. *Antibiotic resistance* is a growing global problem.[24]

The development of *drug tolerance* and both *physical and psychological dependence* are undesirable effects of drugs, all of which can be experienced during the course of chronic drug treatment with legally obtained prescription, non-prescription drugs, controlled drug substances and narcotics. They also can result from recreational experiences with both prescription and non-prescription drugs illegally diverted from properly authorized purchases as well as from using illicit drugs.

Accordingly, due to the importance of these issues for consumer health and safety and to promote more rational drug use by consumers, the following medication-related problems will be discussed in this chapter:

1. *adverse drug reactions*, which include unwanted side-effects of drugs, drug interactions, and drug allergies and idiosyncrasies;

2. *medication errors*, including prescription and prescriber errors, pharmacist dispensing errors, and other health care system errors affecting accurate, safe medication administration; and

3. *antibiotic resistance,* a dangerous worldwide development assisted by over-prescribing, consumer drug non-compliance, and inappropriate non-human uses and handling of human antibiotics.

The following medication-related problems related more closely to consumer-modifiable behavior, knowledge and skills, are discussed in the next chapter:

1. *consumer non-compliance* or non-adherence, which includes intentional and unintentional failures of the patient to follow prescribed and labeled directions, warnings and cautions;

2. *consumer health literacy*, which affects individuals' ability to understand and utilize the health system and health information, including their ability to understand and comply with medication instructions, warnings and cautions; and

3. *drug misuse, overuse and abuse* of prescription and non-prescription drugs, as well as the abuse of illicit drug substances, tobacco and alcohol.

Underlying the prevalence and severity of these medication-related problems are numerous factors. Among the most important of these as relating to consumer responsibility and the ability of consumers to fend better for themselves is adult literacy. In this case this term is an umbrella term for basic literacy and health literacy, which includes drug literacy. Due to the extreme importance of consumer literacy to individuals' safety and effective and affordable medication use, health and drug literacy are discussed in this chapter. It will be seen more clearly that health literacy is inextricably linked to many of the medication-related problems explored in this chapter.

All too often, consumers take these more cavalierly as if they were mere commodities. Following this chapter, hopefully, the reader will better appreciate the basis for the promises and cautions properly associated with the availability and intelligent use of these healing miracles. The reader will learn certain basics about drug actions and reactions in the body, the most important types of drug-related problems, and the prevalence, costs and consequences of the problems related to the use of medication.

The purpose of this chapter is to provide a general understanding of potential risks associated with taking drugs, including man-made problems associated with taking prescription or non-prescription drugs alone or in combination with one another or in combination with nutritional supplements, herbals, alcohol, certain foods or, in special cases, certain diseases or other health conditions; as well as medication errors made by health care providers, care-givers and patients.

At the same time, it is intended to provide assurance of the value and relative safety of FDA-approved drugs when properly used by patients who need them. Accordingly, consumers who need to take legal drugs for a medically useful purpose, as determined by their doctors or other competent health care professional, should feel confident in using them in strict conformance with verbal and written and/or printed instructions accompanying these products.

It is also intended to inform consumers of the consequences of and caution them against ignoring compliance with verbal and printed instructions from prescribers and pharmacists for properly and safely taking, handling and storing these powerful healing and palliative agents. The information in this chapter will provide the basis for achieving these objectives by enabling consumers to apply what they learn to more actively participate on their own health care team with their doctors and other health care providers with more informed questions, observations, ideas, understanding and compliance.

Drug Actions and Reactions in the Body

In order to better understand how most problems associated with taking medications can occur, it is helpful to understand certain basic body and drug characteristics, effects, chemical reactions and interaction mechanisms without needing to hold an advanced degree in chemistry, physiology or pharmacology. A few of these will be explored and explained now.

Whenever foreign substances, including prescription and OTC drugs, are introduced into the body, the body has been programmed to respond in various ways for survival. Unfortunately, the desirable effects of these drugs can be overshadowed by the undesirable effects of a drug.

The desirable effects are usually straightforward and obvious as these are the FDA-approved indications or uses of the drug and as such are clearly listed in the product's approved labeling. Undesirable effects may not be so obvious although clinical studies and practice will indicate the possible and most probable undesirable effects as well as whether they are harmless, although undesirable, or harmful or toxic side effects.

With respect to either desirable or undesirable effects, the active ingredient can affect the body at the local level or systemically or its local or systemic effect may be conferred by the dosage formulation used and its site of application; e.g., externally on the skin versus orally by mouth. Systemic effects occur throughout the body instead of only at the site of application.

Some effects of drugs are more permanent and are referred to as irreversible while most are reversible; that is, upon termination of the administration of a drug, the effect will stop either immediately or over time and the consumer's state of condition will revert to its previous or normal state. In either case, these effects can be immediate or delayed so that they occur only days, weeks or months after taking or discontinuing a drug.

Many times the action of the drug is directly due to the active drug included in the dosage form, while the action of others may be due to one of the breakdown products (metabolites) of the drug after it enters the body and is absorbed. In any case, the mechanisms for drug action in the body follows a general pattern involving the following four basic processes:

1. *Absorption of the active ingredient* from the drug dosage form in the mouth (under the tongue, for example), the stomach, the intestines, the skin to which it is applied, or from the muscle into

which it has been injected, or from any other local site where it may be applied.

2. *Biotransformation of the drug* once it has been absorbed into the body.

3. *Distribution of the active drug* to various body systems' tissues and organs and on to its site or sites of action.

4. *Excretion of the used active drug and/or its breakdown products* via the lungs (breath), kidneys (urine), liver (bile), etc.

During these processes, various kinds of chemical reactions can take place resulting in *allergic reactions* (allergies), *idiosyncratic reactions* (idiosyncrasies) and *interactions* between various chemicals in the body (*drug interactions*). Each of these types of chemical reactions is discussed and illustrated in greater detail later in this chapter.

Complicating matters further is that chemical interactions can take several forms which lead to four basic types of therapeutic effects:

1. *Additive effects*, which means the cumulative pharmacological actions of two ingested drugs increase roughly by the sum of their individual effects;

2. *Antagonism*, where the effects of one drug works against or cancels out some or all of the usual effects of another drug;

3. *Synergistic effects* where the result from the combination of the drugs acting together cause an effect greater than that predicted from the additive separate effects of the individual agents; and

4. *Potentiation*, a term sometimes used interchangeably with synergism, to describe a combined effect which is greater than a mere additive effect.

Some drugs are safer than others and, in general, drugs may be safer for teens and young adults than for younger children and the elderly. However, the benefits of taking any drug must always be weighed against the risks involved in taking it. That is why discussions in this chapter are so important for consumers.

Fortunately, for most people and for most drugs, risks are minimal when taken in accordance with printed label and verbal instructions of prescribing doctors and cautions and warnings from the pharmacist. However, studies indicate that consumers frequently do not follow instructions for taking their medications properly for various reasons. Therefore, risks directly associated with taking individual and combinations of prescription and OTC drugs are discussed along with other factors which can make these risks reality or which can increase their severity when they occur. The problems these risks portend are termed adverse drug reactions (ADRs) or, more broadly, adverse drug events (ADEs).

Adverse Drug Reactions and Events

Adverse reactions to medications (ADRs) constitute a serious health problem because of the sheer number of people affected by them and their sometimes serious, even fatal, nature. Yes, the risks of drug side-effects and other types of ADRs, ranging from mild to lethal, are an unfortunate fact of life for drug consumers with all drug use. That is why it is so important for consumers to be aware of their potential so that their medication experiences can be as trouble-free as possible.

Discussions of ADRs in this chapter cover the unavoidable ones, as well as those which are caused by medication errors and by the failure of consumers to take their medication properly, both of which are more properly referred to as *adverse drug events* (ADEs). An ADE is a medication-related problem defined as "any injury resulting from a medical intervention related to a drug."[25] In this book ADEs are inclusive of both adverse drug reactions which occur during the course of proper drug use and those occurring when medications are taken improperly or not taken when they should have been, whether intentional or unintentional.

According to the World Health Organization (WHO), an adverse drug reaction (ADR) is a more restrictive term related to reactions encountered during appropriate use of a drug. That is, an ADR "is an effect which is obnoxious and unintended and which occurs at doses commonly used in man for prophylaxis, diagnosis or therapy." In other words, adverse drug reactions are undesirable, obnoxious responses to drugs which can be experienced by consumers even when using the drugs properly and in acceptable dose ranges.

Levels of Severity of Adverse Drug Reactions

Adverse drug reactions may be *classified on the basis of their severity as being mild, moderate, severe and lethal.* In accordance with this classification, they can be:

- *Mild* adverse drug reactions, which don't require antidotes or corrective therapy nor prolong hospitalization.

- *Moderate* adverse drug reactions, which require changes in drug therapy, but not necessarily stopping the drug therapy." These may require hospitalization, lengthen hospital stays or require special treatment.

- *Severe adverse drug reactions* are "potentially life-threatening, requiring discontinuation of the drug and specific treatment of the adverse reaction."

- Those ADRs which cause, directly or indirectly, the death of a patient are referred to as *lethal adverse reactions.*

Most often, these effects can be minimized or prevented if the patient provides and the prescriber has available and reviews the complete medical history and list of drugs—both prescription and OTC, including vitamins and minerals and herbal products—being taken by the patient. It is not only smart, but vitally important for patients to make sure that the prescriber and the pharmacist have a complete list of the medications and supplements being taken so these health care professionals can check for any potentially harmful drug interactions. Also any known allergies should be reported.

ADRs vary in their frequency of occurrence and severity, depending on the drugs taken or administered and the gender, age, genetics, and health status of the patient taking the drug. Some drugs are more problem prone than others but it is important to know and remember that all drugs—prescription or non-prescription, legal and illicit—carry some risk of adverse effects when used.

Classes of Adverse Drug Reactions

From a safety standpoint, ADRs either are preventable or non-preventable. The common categories of ADRs are:

1. *Side-effects* are undesirable effects in a given therapeutic situation that are predictable, dose-related pharmacologic or clinical responses occurring within the normal therapeutic dose range.

2. *Drug interactions*, which may be antagonistic, synergistic or additive, occur when two or more drugs are taken by or administered to an individual (drug-drug interactions); or when drugs interact with foods like grapefruit juice or cheese (drug-food interactions); or when drugs interact with a disease or other medical condition a person has and causes or worsens it.

3. *Drug allergies*, which are not dose-related, are common and can cause serious toxicity.

4. *Overdosage* toxicities are predictable, the severity of which is usually dose-related, and occur in doses above the therapeutic range for the particular patient.

5. *Idiosyncracies* are those undesirable reactions to drugs which are unwanted and peculiar; that is, they are not able to be categorized in either of the above four groups.

After a discussion of the scope of potential side-effects, descriptions and examples of the above categories of ADRs will be discussed. The prevalence of preventable ADRs, their costs and consequences are discussed to illustrate the gravity of this problem, their effects on consumer health care costs and quality of life, and other effects on consumer safety and economics. The main focus in this book, however, is on preventable adverse drug reactions, an area where industry, government, the health care system, health care providers and the consumer can make a difference with commitment and vigilance.

Preventable Adverse Drug Reactions

Most attention to ADEs and ADRs have focused on patients in hospitals and other health care facilities and less on outpatient populations. This is understandable since inpatient facilities are better equipped to have collection and reporting systems to capture and report such information. While the FDA runs a voluntary adverse drug reaction reporting system, the collection and reporting of information from ambulatory consumers and health care providers in non-institutional settings can be expected to be more sporadic and less thorough.

More recently, the Consumer Product Safety Commission and the Centers for Disease Control and Prevention have teamed up to jointly provide a means for gathering statistics on adverse drug events as a component of the overall compilation of injuries from all causes in the United States.

While patients of all age groups are susceptible to medication-related problems, ADRs are more likely to be experienced by the very young and elderly consumers than for other age groups for several reasons.

Elderly patients are more likely than younger age groups to be taking multiple medications for both chronic and acute illnesses. As the number of medications taken increases, so does the risk of encountering medication-related problems. These patients are much more likely to experience multiple adverse effects from the drugs they are taking as well as potential interactions between various drug they are taking. They are more likely to encounter multiple drug interactions between two or more drugs they are taking and between drugs and their disease states. They are also more susceptible to serious injury from falls as a result of physical disabilities especially when drug effects alter their senses.

While most adverse drug reactions are not serious, it is estimated that the more serious adverse drug reactions are responsible for 100,000 plus deaths annually and result in some 2 million people being hospitalized or being seriously injured every year.

Adverse drug reactions cause some 700,000 emergency visits to hospitals annually, according to a two-year study the U.S. Centers for Disease

Control and Prevention reported in a recent issue of the Journal of the American Medical Association.[26] According to study findings:

1. One of every six emergency room visits resulted in hospitalization in an inpatient unit or emergency department observation bed.

2. Patients 65 and older were more than twice as likely to be treated in an emergency room for an ADE than younger residents and seven times as likely to be hospitalized as a result.

3. Over a two-year period, some 21,298 ADE cases were reported for an estimated 701,547 patients who visited an emergency room due to ADEs.

4. Most of the hospitalizations from ADEs were due to unintentional overdoses of medication.

    a. Drugs which require outpatient monitoring to prevent acute poisoning were involved in most of the unintentional overdoses. These included diabetes drugs, warfarin (blood thinner), anticonvulsants (for epilepsy), digitalis glycosides (for heart problems), theophylline (for asthma) and lithium (for bipolar disorder).

    b. These drugs were involved in approximately two-thirds (66 percent) of the estimated overdoses requiring hospitalization and about 41.5 percent of all hospitalizations.

    c. For patients 65 years and older, those drugs were involved in 85.4 percent of the estimated overdoses which required emergency room visits, 87 percent of the overdose cases requiring hospitalization and 54.4 percent of all hospitalizations for that age group.

5. Overall, the five most common classes of drugs involved in ADEs were insulin, opioid analgesics, anticoagulants, antibiotics containing amoxicillin, and antihistamines or cold remedies.

6. The five most common classes of drugs involved in hospitalizations from ADEs were anticoagulants (warfarin, etc.); insulin for

diabetes; opioid analgesics for pain; oral hypoglycemic agents for diabetes; and antineoplastic agents for cancer.

7. Sixteen of the 18 medications most commonly reported in an ADE which resulted in an emergency room visit were older drugs which have been in clinical use in the United States for more than 20 years.

On a more positive side, certain drug side-effects are taken advantage of by physicians to achieve patient friendly results; that is, they are used by their prescriber to achieve a desired therapeutic benefit. A common example of this is using the side-effect of drowsiness caused by older antihistamines—anti-allergy drugs—like diphenhydramine (Benadryl) and chlorpheniramine (ChlorTrimeton) to induce sleepiness in patients with sleeping problems or insomnia.

Adverse Drug Reactions in Elderly Persons

Important changes in body structure and function occur as a natural consequence of aging. Elderly patients are more prone to being adversely affected by drugs due to the slower rates of elimination of drugs from their bodies than is the case for younger patients. Consequently, ignoring naturally occurring changes as consumers age can be both costly and deadly.

A drug like erythromycin, an antibiotic, can lead to increased toxic effects if taken or applied by elderly patients taking certain types of drugs like benzodiazepines (for example, lorazepam), calcium channel blockers (for example, amlodipine and nifedipine), cyclosporine, tacrolimus and warfarin, the blood thinner.

Ciprofloxacin, another antibiotic, inhibits the breakdown of the asthma drug theophylline in the body, resulting in a buildup of the drug above therapeutic limits. This causes it to become toxic and increases the risk of seizures in elderly patients.

Also elderly persons may be more prone to suffering impaired memory, developing diabetes mellitus and peptic ulcer when taking both oral corticosteroids (like methylprednisone) and pain killers which are non-steroidal anti-inflammatory agents (NSAIDS, like ibuprofen and aspirin).

First generation antihistamine (anti-allergy) diphenhydramine is more likely to cause cognitive impairment (confusion or memory loss) in elderly patients; and those with reduced kidney function may need dosage reductions by half when taking the newer allergy medication cetirizine (Zyrtec) or methotrexate (MTX).[27]

Costs and Consequences of Adverse Drug Reactions and Events

According to findings of the IOM in its 2006 report on medication errors, preventable adverse events with prescription drugs occur at the rate of 1.5 million each year, resulting in excess of 7,000 deaths.[28] Of these, an estimated 380,000 to 450,000 occurring in hospitals annually are deemed to be preventable. Some 800,000 preventable ADRs are estimated to occur in long-term care and some 530,000 ADRs are estimated to occur in Medicare only ambulatory care patients.

These translated into estimated costs of $6,000 per preventable ADR in hospitals for a total annual cost of some $3.5 billion in 2006 dollars and $887 million for preventable ADRs in ambulatory care in 2006 dollars. Comparable figures were not available for preventable ADRs occurring in long-term care facilities.

Drug Interactions Are Serious Business

All pharmaceuticals and foods are comprised of chemicals and, as such, possess the potential to react chemically or physically with one another. This applies to prescription drugs, non-prescription drugs, foods, beverages like alcohol, individual vitamins and minerals, multiple vitamin-mineral supplements, herbs and other natural products, as well as with environmental chemicals.

A drug interaction is simply the interaction of a drug with another drug or other substance. However, the steps involved in clinically significant drug interactions in the body consist of more complex processes involving various enzyme systems. The outcome of an interaction may be clinically insignificant, favorable or unfavorable. From the standpoint of consumer safety and drug effectiveness, drug interactions are red flags in prescription drug therapy since their results can be serious, even fatal.

Types of Drug Interactions

In the practical sense, a drug interaction can be a *drug-drug interaction*, meaning that it involves the interaction between two or more drugs. These can be prescription or non-prescription drug products. Drugs can also interact with certain disease conditions to cause or exacerbate a person's health. Drug interactions with patient disease states are referred to as *drug-disease interactions*. Interactions also can occur between a drug and a food substance (*drug-food interactions*). *Drug-substance interactions* involve the interaction between drugs and alcohol, tobacco, herbs or an environmental chemical.

While many drug interactions are unremarkable from a clinical perspective, pharmacists and physicians are trained, experienced and use their knowledge along with computerized programs to screen for bothersome as well as serious drug interactions between prescribed medications, OTC drugs and dietary and herbal supplements the consumer has informed them that he or she is taking. That is why consumers must make certain that these doctors and pharmacists have in their records all medications they are taking, including dietary and herbal supplements, and any other natural products.

Interaction of drugs with other drugs

Since there are thousands of drugs, our discussion will center on some of the most common drug-drug interactions to illustrate the potential problems associated with this type adverse drug effect as well as to indicate ways to prevent one type of preventable medical errors—ADRs.

Drug-drug interactions can occur via several mechanisms at different sites before and after the drugs enter the body:

1. Drugs interactions can occur even before drugs enter the body due to the incompatibility of ingredients in a drug formulation, or at any point in the process of absorption, distribution metabolism, and elimination.

2. Drugs can bind to each other in the gastrointestinal (GI) tract, preventing absorption of the drugs and thereby limiting their availability for their intended systemic activity.

3. Many clinically important interactions take place in the liver and GI tract because other medicines induce or inhibit their metabolism. When this changes their rates of drug metabolism, it affects the outcome of clinical response to these drugs.

4. Other interactions occur through a number of elaborate, more complex mechanisms. These mechanisms include competition at drug transporters and interactions at the specific level of drug action in the body. An example of this would be an action caused by combining two different drugs, each of which slows the heart rate but by different mechanisms.

Interaction of drugs with disease conditions

The term *drug-disease interactions* refers to the avoidance of specific drugs in patients who have certain disease conditions which the drugs may worsen if taken. A list of examples of diseases and prescription drugs which can exacerbate are too voluminous to include in this book. However, just about every major organ system in the body, when diseased or deficient, can be adversely affected by specific drugs. A small sample of the disease conditions most commonly the focus of drug-disease interactions includes asthma, congestive heart disease, diabetes, glaucoma, hypertension or high blood pressure, thyroid problems, and kidney and liver disease.

Interaction of drugs with foods and dietary supplements

Foods and foodstuffs with which ordinary prescription drugs and some OTC drugs are more likely to interact include the following:

- *Grapefruit juice*, grapefruit and a limited number of other citrus products can interact with a large and widening group of prescription drugs due to a substance found in the fruits' peelings.

- *Iron supplements* chelate (combine with to form insoluble compounds) with some antibiotics and thereby render them ineffective.

- *Salt substitutes* can raise the concentrations of certain drugs to toxic levels in the blood;

- *Cheese* interacts with some depressants which are monoamine oxidase inhibitors (MAOIs), such as Marplan (isocarboxazid), Nardil (phenelzine), and Parnate (tranylcypromine).

- *Beverages, especially those with alcohol,* can multiply the depressant effects of some drugs and nullify the effects of others, especially some antibiotics.

Foods can interact with drugs in a variety of ways with the result that the effectiveness of the drug is reduced or the absorption of food nutrients is diminished. Vitamin and herbal supplements may also interact with prescribed or non-prescription medication with unwanted consequences. The most likely results of food interactions with a drug are adverse drug reactions such as:

- the actions of the drug can be speeded up or slowed down;

- the absorption of vitamins and minerals may be impaired;

- the appetite may be stimulated or suppressed;

- the use of nutrients in the body may be altered; or

- herbal components may interact with anesthesia, certain high blood pressure medication and anticoagulants like warfarin.

Grapefruit juice has a greater likelihood of interacting with drugs in the following classes, although the list continues to grow:

- antiarrhythmics, including amiodarone (Cordarone), disopyramide (Norpace) and quinidine.

- calcium channel blockers felodipine (Plendil), nicardipine (Cardene), nimodipine (Nimotop), and nisoldipine (Sular).

- statins for cholesterol disorders, including atorvastatin (Lipitor), lovastatin (Mevacor) and simvastatin (Zocor).

- a protease inhibitor used in the treatment of HIV/AIDS, saquinavir (Fortovase).

Fortunately, there are alternative drug therapies in each of the above cases. However, if one decides to remove grapefruit juice from his/her diet, it greatly broadens the variety of drug therapies that are available to treat that drug consumer. Among others, the management of drug-grapefruit interactions is discussed in the August 15, 2006 issue of American Family Physician.[29]

Iron supplements like ferrous sulfate taken along with the *tetracycline antibiotics* (oxytetracycline, tetracycline, etc.) form an insoluble complex in the stomach, preventing absorption of the antibiotic. This, of course, effectively cancels the desired action of the antibiotic which, to be effective against infections, must be absorbed into the blood stream.

Ordinary table salt is comprised of sodium chloride. *Salt substitutes* are edible products designed to taste as close as possible like ordinary table salt. LoSalt is a well known salt substitute brand with a ratio of two parts potassium chloride to one part sodium chloride. Such products are often used to lower sodium intake because of the serious effects excessive amounts of sodium in the body can have on blood pressure and associated heart conditions. Medical advice is advisable for anyone choosing to use salt substitutes in their diet.

Salt substitutes most commonly contain potassium chloride which, when eaten, increase the concentration of potassium in the body. However, there are also various botanical/herbal blends designed to quench one's thirst for salt. Excessive quantities of potassium in the body result in a condition called hyperkalemia, a potentially fatal condition. The excretion of potassium from the body can also be decreased by the presence of certain diseases, such as kidney or heart failure or diabetes, and medications such as aldactone, amiloride, eplerenone, and spironolactone.

Certain water pills or diuretics, like hydrochlorothiazide, used to control blood pressure, are incompatible with the use of *salt substitutes*, as this combination results in increased blood levels of the mineral potassium. The effects can be from mild, like nausea and vomiting, to more serious muscle weakness and cardiac arrest.

These are but a few examples to illustrate why it is important for drug consumers to continue to read their drug labels and drug information pamphlets to become more drug literate in as many ways as possible for their own safety and welfare and for their families.

Monitoring and Reducing Drug Interactions

Both the prescribing physician and pharmacist are responsible for monitoring a patient's medication to prevent harmful interactions. However, in order for them to carry out this responsibility, the patient is responsible for providing information on drugs and other substances being taken on a regular basis. Only in this way is it possible for the doctor and pharmacist to have in a patient's medication record every prescription, especially those prescribed by other doctors or dispensed at other pharmacies, as well as all OTC drugs, dietary and herbal supplements, and natural products they are taking on a regular basis.

Drug consumers play an important role in reducing and minimizing unwanted drug interactions for their own health and safety and for members of their own families. Several of the most important steps to be taken by drug smart consumers in this direction are:

1. Maintain a current, complete and accurate list of all prescription and non-prescription drugs and dietary and herbal supplements they or their family members are taking.

2. At each office visit, tell their doctors all prescription and non-prescription drugs and dietary and herbal supplements they are taking.

3. At each pharmacy visit, tell the pharmacist all prescription drugs they have that were filled at another pharmacy and non-prescription drugs and dietary and herbal supplements they are taking.

4. Update their medication list each time a new prescription drug is filled, or another OTC drug or dietary or herbal supplement is purchased for use.

5. Read the drug information pamphlets accompanying their filled prescriptions and ask their doctor or pharmacist questions.

6. Read the caution and warning labels on all prescriptions filled and non-prescription products used.

Drug Allergies and Idiosyncrasies

Allergic reactions and hyper-sensitivities can be serious business. A drug sensitivity is defined as an unusual reaction to a drug not involving the immune system while an allergic reaction is the result of the immune system's response to the presence of the drug. An allergic reaction is the result of one's body incorrectly reacting to a substance which is harmless to most people.

In an allergic reaction, when one is first exposed to a medication, the immune system is triggered and becomes sensitized to it. A second exposure to the drug causes the immune system to respond to it as a foreign body by producing antibodies and histamine. As a result, the body exhibits symptoms such as skin rashes, hives, itching skin or eyes, wheezing, swelling of the tongue, swelling of the lips or face, or more serious anaphylaxis, which can be life-threatening.

An *anaphylactic reaction* will usually occur within the first few minutes after taking the first dose of the medication. However, it can occur several days after the drug is first taken.

Symptoms of an anaphylactic reaction include difficult breathing with wheezing or hoarseness; hives over various parts of the body; fainting or light-headedness; heart palpitations; abdominal pain or cramping; confusion; diarrhea; dizziness; nausea and vomiting; and rapid pulse. In such a case, immediate medical attention is called for by calling one's physician or the local emergency number, 911, or being rushed to a hospital emergency room for treatment.

The good news is that allergic drug reactions occur rarely and most are minor. The bad news is that drug allergies are not curable and, in the more severe cases, can result in anaphylactic shock and death, if immediate medical attention is not gotten. However, no drug is completely free of allergy potential.

The drug types responsible for drug allergies most frequently include the following classes: anesthetics, antibiotics, anticonvulsants, barbiturates, heart medications, heavy metals, hyperthyroidism medications, laxatives, organ extracts, sleeping pills, tranquilizers and vaccines. However, while allergic drug reactions will vary in type and severity, the chemical class of drugs from which patients are most likely to have such a reaction are the antibiotics. For illustration, only a few allergy prone drug classes will be discussed since a listing of those from each of the aforementioned classes would be to extensive for the purposes of this book.

Some of the most common antibiotics responsible for allergic reactions are the penicillins (amoxicillin, carbenicillin, penicillin V potassium, etc.), sulfa drugs (sulfamethazole, etc.), chloramphenicol, neomycin, and streptomycin.

Penicillin and cephalosporins

Examples of drugs in the penicillin group are *amoxicillin, ampicillin, carbenicillin and dicloxacillin.* The drug Augmentin® is comprised of amoxicillin and another ingredient. Unfortunately, a sizeable percentage of the population (ca. 10 percent) believe they have a penicillin allergy. Because patients often refer to any unpleasant feeling after taking a penicillin as an allergy, it is probable that many fewer people have experienced a true penicillin allergy than the numbers reported. Although a serious allergic reaction can be life-threatening, most reactions are manifested in itchy skin and eyes or hives.

Some people with a penicillin allergy may also experience allergies with members of another antibiotic family, the cephalosporins: *cefaclor, cefadroxil, cefepime, cefprozil, cephradine, cephalexin* and *cephotetan.* Needless to say, this applies equally to brand-name and generic drug products containing these active drugs.

Sulfonamides and sulfur-containing drugs

The sulfonamide group includes drugs like *sulfamethazole* and *sulfathiadiazine*, with *sulfasalazine* being the most widely used. Persons allergic to sulfa drugs will also need to steer clear of other products containing these active ingredients alone or in combination with other drugs. These include the branded drugs Bactrim, Septra and Pediazole.

Consumers who have experienced allergies with a sulfa drug should not take any other sulfa or sulfonamide drug and should make certain that their physician and pharmacist know about their sulfa allergy. Since generic drug products contain the same active ingredients as the brand-name products, any allergic reactions to sulfa drugs apply equally to both.

Anticonvulsants/anti-epileptic drugs

Anticonvulsants are used to treat episodes of epilepsy. Anticonvulsants which have been shown to cause allergic reactions more frequently are Depakene (valproic acid), Dilantin (phenytoin), Lamictal (lamotrigine), and Tegretol (carbamazepine).

Human insulin versus insulin from animal sources

Before the discovery of human insulin, pork and beef insulin caused allergies in many individuals. These two types of animal insulin were widely used worldwide and accounted for more frequent occurrences of allergic reactions. Now all insulin approved for marketing in the United States is of the human variety, not from animal sources, and produced with a high technology referred to as recombinant DNA technology. However, there can be no absolute guarantee of freedom from allergies even for these products even with such an improved allergy profile.

The following major brands of human insulin are commercially available in the United States: lispro or Humalog, Novolin, aspart or Novolog, glargine or Lantus, and Exubera, the first and only non-injectable insulin, which is administered by inhalation.

Responding to drug allergies

It is always safest to consult a doctor or pharmacist when one suspects that an allergy has occurred. It is essential and absolutely necessary to seek professional attention and emergency assistance when a serious allergic reaction is experienced, as this can be life-threatening.

For allergies in general, the following treatment strategies are generally recommended:

1. The most effective strategy always is to avoid the offending substance, when this is possible.

2. For mild reactions, antihistamines are commonly used to control the itching, rashes or hives. OTC ointments, creams and sprays are often adequate when applied to the local areas when these areas are not too widespread. Otherwise, oral antihistamines are often used. Since these are primarily non-prescription medications, the doctor or pharmacist will make appropriate recommendations.

3. More severe symptoms, or anaphylaxis, require more immediate emergency medical attention. In such cases, the use of corticosteroids, much more powerful drugs than the antihistamines, may be necessary; an epinephrine injection may be necessary; or in the case of moderate wheezing or asthma-like coughs, the use of a bronchodilator (albuterol, for example) may be necessary. Physicians will determine the appropriate medications to use for this purpose.

Medication Errors Are a Major Problem

Medication errors represent a troublesome and potentially life-threatening group of preventable medication-related problems. These can be caused by prescribers, nurses who administer prescription drugs to patients in hospitals and other health care institutions, dispensers and care-givers. This section focuses on medication errors caused by:

1. prescribing physicians and other prescribers;

2.   nurses who administer these medications to patients in hospitals and nursing homes pursuant to doctors orders; and

3.   pharmacists who receive, interpret and dispense the patient's prescriptions.

The formal definition of the term, which is useful for national uniformity in monitoring, researching and applying knowledge about this group of problems, has been developed by a national group of health care organizations under the aegis of the United States Pharmacopeia, the official compendium. This working group is the National Coordinating Council for Medication Error Reporting and Prevention (NCCMERP).

According to the NCCMERP, this definition is:[30]

> *"A medication error is a preventable event that may cause or lead to inappropriate medication use or patient harm while the medication is in the control of the health care professional, patient, or consumer. Such events may be related to professional practice health care products, procedures, and systems, including prescribing; order communication; product labeling, packaging, and nomenclature; compounding; dispensing; distribution; administration; education; monitoring; and use."*

If one considers that medication errors include all mistakes involving prescription drugs, non-prescription drugs, and dietary and herbal supplements, it may be easier to understand that the statistical data represent only the tip of the iceberg due to an unavoidable incompleteness of data. Reported medication errors occurring in the retail or community pharmacy settings remain much less complete than data captured in institutional settings in large part due to procedural limitations. Even so, the reported findings are stunning in terms of the prevalence of medication errors in the U.S. health care system.

Preventable medication errors take their toll in human lives, lost productivity and increased health care costs for all concerned. While medication errors have long been a serious health care problem, a 1999 report by the Institute of Medicine (IOM) initially focused national attention on the problem and the added costs associated with *medical errors* in the U.S.

health care system. These studies estimated that at least 400,000 preventable injuries and deaths occurred in hospitals annually while some 800,000 were estimated to occur in nursing homes and other long-term care facilities annually.

IOM's most recent report, which focused on preventing *medication errors*, showed that these errors either sicken, injure or kill at least 1.5 million people in U.S. hospitals every year. Such errors include medication errors caused either directly or indirectly in prescribing, preparation, labeling, dispensing and/or the administration of drugs whether by doctors, physician assistants, nurses, or dispensing pharmacists. Treating drug-related injuries caused by medication errors occurring in hospitals alone is conservatively estimated to cost $3.5 billion annually.[31]

Prescribing and Prescription-Based Errors

According to an article in Family Practice Management, a journal of the American Association of Family Physicians, the seemingly innocuous disruptions and distractions that affect practitioners in all professional fields also take its toll in prescribing errors. As most consumers are aware to an extent, their encounters with physicians are characterized by increasing competing demands on the doctors' limited time and attention. In fact, an earlier group of researchers, Leappe, L., et al.,[32] found that about three-quarters of transcription errors could be traced back to distractions. The previous authors note that:

> "Distortions that occur when prescriptions are created or deciphered may lead to erroneous substitutions of whole medication regimens and generate severe errors. Because many distortions arise from illegibility and misunderstood translations of symbols or abbreviations, they are also some of the most remediable sources of medical errors."[33]

Recommendations for physicians writing prescriptions to minimize errors and maximize patient safety are summarized in the following 14 tips *(in italicized* letters) for preventing medication errors. Some of these have been modified slightly or expanded upon (in regular type) to clarify the information for consumers to help them better appreciate the prescriber's need to focus during the prescription writing process near the end of their visits in order to minimize errors.

1. *Limit each prescription to one medication.*

2. *Circle the* (the physician's) *name when using preprinted prescription pads* with more than one doctor's name imprinted on them.

3. *Approach medication name with caution.* There are many sound-alike (when spoken) and look-alike (when printed or handwritten) drugs. Misinterpretation of these types of drugs, especially when hand-written, is a common cause of medication errors.

4. *Eliminate drug abbreviations.* Thousands of drugs, drugs, dosage quantities, dosage regimens, medication strengths make it easy for transcription errors to occur when technicians or pharmacists interpret and transfer information from the doctor's prescription into the computer in preparation for filling.

5. *Use metric measures for dosages (mg, mcg, Gm, ml, for instance),* rather than outdated measures, such as apothecary measures (dram, oz or ounce, etc.)

6. *Add the patient's age or weight to the prescription.* This is an important clarification for patients of all ages but particularly for prescriptions for infants, children or seniors as it enables pharmacists to verify safe dosing and potential adverse drug effects.

7. *Avoid writing "as directed" on prescriptions.* These non-specific directions for a patient's dosing prevent the pharmacist from making sure that a patient clearly understands how their medication is to be taken.

8. *Eliminate abbreviations in routes of abbreviation.* Many commonly used Latin abbreviations can be misinterpreted by the pharmacist when written illegibly. For instance, the abbreviations for daily (q.d.), four times daily (q.i.d.) and every other day (q.o.d.) can easily be misinterpreted on illegible prescriptions when cursive

writing is used. These are but a few of the many abbreviation combinations which fall into this category.

9. *Specify the duration of therapy.* Certain medications need to be titrated periodically and for effectiveness and safety reasons, the physician must maintain tighter control over the dispensing and use of dosages of these medications. The number of refills should be clearly indicated and written out for clarity.

10. *Prescribe specific quantities rather than dispensing for time periods.* In order to properly monitor patients' drug use, counsel them on overuse, and schedule authorized refills, pharmacists must have a basis for determining the days' supply for each prescription. Prescriptions written with unclear or incomplete directions or "as directed" directions fail to provide this information.

11. *Remain cognizant of lethal doses of medications.*

12. *Specify the indication* (drug use) *on the prescription.* Many medications can be prescribed for more than one type of illness or condition and the indication can assist the pharmacist in counseling the patient more appropriately about the prescription. Often, the pharmacist is unable to tell the patient the specific reason the drug has been prescribed for them because of possible multiple uses of the drug.

13. *Write additional instructions about medication side-effects.* Most pharmacists, however, provide printed drug information along with a prescription, which includes the more common and less common side-effects which can be expected from taking a prescription medication.

14. *Report and review all errors* for use in continuous quality improvement.

Dispensing Errors and Omissions

Pharmacists are some of the most meticulous and accurate health care professionals, in spite of the "mega multi-tasking" they perform under unrelenting pressures from multiple patients and other consumers, prescribers, insurers and pharmacy benefit managers, while simultaneously receiving, validating, filling, verifying, drug utilization review, patient profile monitoring, and counseling patients on drug use. Since safe and effective drug therapy is their priority, their accuracy in filling prescriptions in general is phenomenal. Even so, too many dispensing errors still occur.

Dispensing errors are those caused by the pharmacist or other member of the pharmacy's staff and relate to the specific patient for which a medication is prescribed, the drug prescribed, the drug dose and dosage form prescribed, directions for taking the medication, quantity of medication dispensed, reviewing a patient's medication profile, dispensing container and closure, and any specific cautions and warnings needed to ensure safe and effective use of the prescribed drug.

The following is only a partial list of the kinds of errors which can occur in the pharmacy when processing and dispensing prescriptions:

1. Incomplete, inaccurate or improper label instructions;

2. Incorrect drug dispensed, often due to illegibly written prescriptions or from confusion of look-alike, sound-alike drug names;

3. Incorrect drug dosage form and/or strength dispensed;

4. Incorrect quantity of drug dispensed;

5. Improper mixing of liquid or semi-solid products;

6. Dispensing without mixing liquid or semi-solid products;

7. Dispensing expired drug products;

8. Failure to check for drug interactions;

9. Failure to check for drug allergies;

10. Lack of cautions or incorrect cautions on label;

11. Dispensing in wrong type container and/or closure;

12. In hospitals, improper distribution in the pharmacy, to satellite locations and to nursing stations;

13. Failure to monitor a patient's drug use; and

14. Lack of patient education in printed and/or verbal form.

In order to better protect consumers against dispensing errors, the boards of pharmacy in states like California, Florida, Iowa and Texas have instituted requirements for community pharmacies, as well as hospitals and other institutional pharmacies where such requirements have long been required, to document dispensing errors and implement continuous quality improvement (CQI) programs. Such programs are proactive activities aimed at continuously reducing the risk of medication errors by pharmacists in order to maximize patient safety in using prescription drugs.

CQI programs being instituted among the state boards of pharmacy can provide a mechanism for the capture and dissemination of statistical information on pharmacy-based dispensing errors as well. In order for these to be most meaningful, however, there will need to be more standardization of the various dispensing error reporting systems among the states.

Medication Error Prevention and Reporting Programs

Because of the importance of the issue of drug safety, two national medication error reporting systems are in operation in the United States. Their primary interest is promoting and protecting patient safety in the use of drugs and related health care products by monitoring and preventing medication errors. Each of these programs encourages reporting of medication errors and provides forms to consumers, health care professionals and drug manufacturers to facilitate the reporting. These reporting systems, their sponsors and toll-free contact numbers are:

- The *MedWatch Reporting Program* at the Food and Drug Administration, 1-800-FDA-1088. Consumers are encouraged to report suspected serious reactions to drugs either through their physicians or directly via forms available over the Internet. Most health care providers should but are not required to report these to the FDA.

- The *Medication Error Reporting Program* (MERP) under the auspices of the U.S. Pharmacopeia (U.S.P.), the official national drug compendium, which forwards medication error reports it receives to the FDA. Its contact number is 1-800-233-7767. In this program, the U.S.P. MERP is acting in partnership with the FDA to capture the following types of errors reported by pharmacists, nurses, physicians, and other health care professionals: misinterpretations, miscalculations, mis-administrations, difficulty in interpreting handwritten orders, and misunderstanding of verbal physician orders.

- The *Institute for Safe Medication Practices* (ISMP), which also forwards medication error reports it receives to the FDA.

The FDA and the ISMP have joined forces in a national education program to eliminate the use of ambiguous medical abbreviations used in prescription writing, which frequently lead to mistakes in interpretation and result in harm to patients. The ISMP has published a list of error-prone abbreviations, symbols and dose designations which has been recommended by the FDA as part of the joint campaign.

Another national organization which focuses on medication error prevention is the *National Coordinating Council for Medication Error Reporting and Prevention* (NCCMERP). Its mission is "to maximize the safe use of medications and to increase awareness of medication errors through open communication, increased reporting and prevention of medication error prevention strategies." Its recommendations serve as a basic standard for the various health care professions organizations to use in minimizing medication errors even with increasing use of technology to assist in this effort. One of the important contributions to this effort is the NCCMERP Index for Categorizing Medication Errors which distinguishes between the levels of severity of medications errors committed.

According to the NCCMERP Index, medication errors are classified as:[34]

1. No error or the absence of an error.
2. Errors which result in no harm to the patient.
3. Errors which result in harm to the patient.

4.   Errors which result in the death of the patient.

Uniform usage of this classification scheme provides a level of standardization in the reporting and comparison of medication errors among the various types of health care institutions as well as globally to describe the nature and extent of the problem and the degree to which there may be improvement or decrease in the quality of health care as may be indicated by individual and aggregate records.

Antibiotic and Antimicrobial Resistance

More than 50 million unnecessary antibiotic prescriptions are written each year for patients outside hospitals, according to the U.S. Centers for Disease Control and Prevention (CDC). Also, according to CDC, it is clearly established that virtually all important bacterial infections in the United States and worldwide have become resistant to the most effective antibiotics available for their control.

Annually, some 2 million patients get infections while in a hospital. Of these, more than 90,000 patients die from such infections. Such antibiotic resistant organisms are adept at spreading from these institutions into the general population, further complicating the global problem of antibiotic resistance.

Hospital-acquired infections, referred to as nosocomial infections, continue to spiral out of control throughout our nation's facilities despite frantic efforts to control and prevent them. These infections occur as a result of treatment in a hospital or other institutional environment where large numbers of people are sick whose immune systems are already weakened. The movement of medical staff from patient to patient facilitates the spreading of antibiotic-resistant disease-causing bacteria. The body's natural protective barriers are routinely bypassed in many medical procedures. The routine use of antibiotics and other anti-microbial agents in hospitals in an effort to control and prevent infections further contribute to this problem by promoting mutation of surviving infectious bacteria.

The problem of bacterial resistance to present day antibiotics is, however, much more widespread in hospitals and nursing homes where its effects are much more concentrated. It permeates all segments of the health care system both here and abroad and respects no borders, geographic or otherwise.

To help the reader to understand the problem of antibiotic and antimicrobial resistance, the following topics will be discussed: what a microbe or microorganism is; the most important types of infectious microorganisms; what antibiotic resistance is and what causes it; how antibiotic resistance develops; and the costs and consequences of this problem to society.

Bacteria, Viruses and Infectious Disease

Infectious disease was a major killer before the discovery of the first antibiotic penicillin and the subsequent discovery and synthesis of other antibiotics. The discovery of penicillin and subsequent antibiotics have since saved, improved and extended countless millions of lives worldwide. Bacterial resistance is a natural obstacle in the pathway of continued medical progress in the war on infectious disease.

While microorganisms are virtually everywhere, including inside the human body, under normal circumstances, only certain types and varieties cause infections in man. The organisms which cause infections most commonly in humans and animals are bacteria, viruses and fungi.

Bacteria, the most frequent offenders in infectious disease, are one-cell organisms which are ever present in and on our bodies and on about everything we touch. Many bacteria are harmless and many are actually helpful, especially those which normally inhabit the GI tract of humans and other animals. On the other hand, infectious bacteria cause such familiar diseases as strep throat, bacterial pneumonia, gonorrhea, septicemia, wound infections and a multitude of other mild to severe and serious infections.

Viruses are disease causing microorganisms even smaller than bacteria and can only survive inside body cells. These are the notorious offenders in most colds, bronchitis, influenza, herpes, HIV/AIDS, measles, pneumonia, viral meningitis, viral gastroenteritis, SARS and a multitude of other diseases.

Understanding Antibiotics and Other Antimicrobial Drugs

Antibiotics are drug substances which fight infections caused by bacteria. Antibiotics are necessary, life-saving tools for saving and improving the quality of life of millions by preventing and treating a wide variety of mild to life-threatening infections in all age groups on every continent. Since the initial discovery of the first antibiotic penicillin in 1927 and its first use in the 1940s, there has been continuous discovery of newer and more potent antibiotics over the years, as it was found that over time each antibiotic seems to gradually lose its effectiveness in preventing, treating and subduing the infections against which it was most effective previously.

Bacterial infections are counteracted or prevented by using antibiotics, antimicrobials or anti-infective agents, which may be natural (the original penicillin) or synthetic drugs or chemicals.

Antibiotics kill or control the spread of microorganisms by a variety of mechanisms, although their general effects are basically two and they are classified by their general effects into two groups.

1. *Bactericidal antibiotics* work by actually killing the microorganisms. In order to accomplish this, the antibiotic must be taken or applied to the body in concentrations large enough to kill all of the organisms.

2. *Bacteriostatic antibiotics* inhibit the growth of the microorganisms. This type of antibiotic similarly must be present in or on applicable body parts in sufficient concentrations that it prevents the growth, reproduction and spread of the offending bacteria.

While newer antibiotics being developed may target other sites of action, the ones most commonly in use today work through the following mechanisms to prevent, cure or treat bacterial infections:

1. Prevent or inhibit synthesis of the bacterial cell wall (example, penicillin);

2. Injure, damage or destroy plasma membranes inside bacteria (e.g., polymyxin B);

3. Inhibit the synthesis of nucleic acid, a necessary component for bacterial life (examples are rifampin and the quinolone antibiotics);

4. Competitively inhibit enzyme activity, a necessary activity for growth and basic processes, in bacterial cells (antimetabolites, including sulfanilamide); and

5. Inhibit the synthesis of protein, also a necessary requirement for all living organisms (chloramphenicol, erythromycin, streptomycin and tetracycline are examples).

Viral infections must be treated with *antiviral drugs*, depending on the infection-causing virus, including such drugs as oseltamivir (Tamiflu) for treating the flu to HIV/AIDS drugs like zidovudine (Retrovir), crixivan (Indinavir) and saquinavir (Invirase, Fortovase), to valocyclovir (Valtrex) for treating herpes to drugs like amantadine (Symmetrel) for influenza and interferon for the treatment of other viral diseases. These drugs are not antibiotics and, therefore, are not effective in the treatment of bacterial infections.

Antifungals are drugs that are effective in the treatment of fungus infections. Such fungus infections include athlete's feet (tinea pedis), usually treated with OTC products like miconazole, clotrimazole or tolnaftate; vaginitis, treated with drugs like clotrimazole, miconazole and terconazole, some of which require prescriptions while others can be purchased over-the-counter; and nail infections (onychomycosis), treatable by an antifungal drug like terbinafine (Lamisil) or similar agents. Some antifungal drugs are effective when taken orally or applied topically to the skin of the target area.

Causes of Antibiotic/Antibacterial Resistance

The causes of antibiotic resistance include man's inappropriate use of antibiotics in humans and in farm and other animals.

Infectious microorganisms called bacteria adapt to sub-lethal doses of an antibiotic by a process called mutation, thereby becoming immune to antibiotics which were previously effective against them when administered in therapeutic doses. Once bacteria have mutated in the presence of sub-lethal doses of a particular antibiotic, it is no longer effective in destroying or inhibiting the growth of these germs, many of which cause serious infections like gonorrhea, bacterial pneumonia, tuberculosis, and a host of other deadly infectious diseases. Mutations occur by several pathways, including replication and the transfer of protective cellular material between different types of bacteria, thereby accelerating the spread of antibiotic resistance to other bacteria.

The causes of antibiotic and antimicrobial resistance are multi-dimensional. The most basic survival mechanism for bacteria is their skill at mutating rapidly in the presence of ineffective doses of an antibiotic. Unfortunately, they are aided in this by man's callous and unintentional contributions to environmental conditions which favor the proliferation of mutated varieties of these microbes.

Life-threatening antibiotic resistance is rooted in multiple causes. *Over-prescribing of antibiotics* for health conditions against which antibiotics are ineffective is one of these. Doctors prescribe tens of millions of antibiotics for viral infections annually due primarily to the following reasons most frequently cited by doctors:

- Diagnostic uncertainty due to an inability to obtain immediate feedback from testing;
- Time pressure on doctors; and
- Patient pressure on doctors to prescribe antibiotics.

Patient demands for antibiotics for conditions for which they are not effective and their insistence on taking antibiotics for viral infections, such as bronchitis, common colds, cough and the flu, promotes antibiotic resistance among previously sensitive disease-causing bacteria. Further consumer *non-adherence* to prescribed antibiotic dosage regimens and failure to take their antibiotics for the prescribed duration of therapy (*non-persistence*), add to this problem. A large percentage of patients stop taking their antibiotics once they feel better, against the advice of their doctors, even though they have not completed their prescribed course of therapy.

Saving these left-over antibiotics for future use continues the upward spiral in antibiotic resistance.

Improper disposal of unused antibiotics is another contributing factor with unquantifiable impact on environmental effects in the growing epidemic of bacterial resistance.

The problem is further exacerbated by *the use of human antibiotics or those in the same chemical class as human antibiotics for non-medical agricultural purposes*, such as for growth promotion in farm animals, a largely preventable practice. Such uses facilitate the development of resistant bacteria in farm animals and various ways and proliferate the spread into the human population through bacterial mutation, replication and cellular material interchange.

It was just in March of 2007 that the FDA ignored the advice of its own Veterinary Medical Advisory Committee to approve a new drug, cefquinome, for treatment of animals. Cefquinome is a fourth-generation cephalosporin, an important class of human antibiotics used in serious infections, including serious GI infections in children and bacterial meningitis. Such an approval represents a dangerous threat to human health and must be opposed with full vigor.

Costs and Consequences of Antibiotic Resistance

Important reasons that failure of antibiotic treatment due to bacterial resistance are many, all of which contribute to increased health care costs for the consumer individually and for the public health care system. As a result of treatment failures caused by antibiotic resistance:

1. Infections last longer than they would if the antibiotic were effective;

2. More doctor visits are required for treatment;

3. Hospital stays are extended for trying alternate antibiotics;

4. More expensive drugs are required, since newer drugs will usually be required;

5. More toxic drugs usually are required;

6.   Infections can spread more quickly; and

7.   Infections can cause death to millions.

Slowing the Spread of Antibiotic Resistant Germs

There are several ways consumers can contribute to minimizing the spread of antibiotic resistance. This important task requires the active participation of drug consumers as well as their doctors, other health care professionals, government regulatory agencies, other government health agencies and patient advocacy groups, in order for the effort to succeed.

Consumers can do a great deal to slow the spread of super bugs resulting from the spread of bacterial resistance. Among the most significant steps average consumers can take to make a big difference with this serious problem are the following:

1.   A first step for consumers is to realize and *accept the fact that antibiotics do not kill viruses.* The failure of a vast majority of consumers to accept this fact, even from their doctors or pharmacists, is a gigantic problem which ultimately works against their receiving favorable treatment outcomes from antibiotic therapy. It is a dangerous, self-defeating consumer behavior.

2.   After acceptance and adoption of Step 1, *consumers must learn not to press doctors to prescribe antibiotics for them when doctors hesitate to do so.* Doctors are trained in medicine and know how to distinguish between common types of upper respiratory infections. In some cases, it may require that a physician prepare a throat culture to determine the exact infectious organism causing the patient's upper respiratory condition. However, growing the culture to make an adequate identification of the specific causative agent cannot be done in a matter of a few hours, not to mention during the course of an office visit. Consumers, therefore, must learn to accept established fact that viruses, rather than bacteria, are responsible for most of the common colds, influenza (the flu); most coughs and bronchitis; sore throats, except "strep" throat; or runny noses.

3. *Consumers must accept* the official guidance from the nation's premier disease prevention and control agency, the CDC, *that taking antibiotics in these situations provides a false sense of security* as this will not prevent them from passing their virus infections on to others.

4. Finally, consumers must learn and accept that each antibiotic taken is only effective against specific infection-causing bacteria. That is the main reason that different antibiotics must be prescribed for different infections.

In line with the above suggested steps and CDC recommendations, consumers can help most in the campaign to slow the spread of antibiotic resistance if they will:

- Take antibiotics exactly as prescribed.

- Take all antibiotic pills for the prescribed course of treatment, even if they feel better before they've taken all of the pills, unless otherwise directed by one's prescriber.

- Throw out any left-over antibiotic medication after finishing the prescribed course of therapy, if any are left. (Discarding these improperly, however, can also add to rather than help reduce the problem.)

- Wash one's hands frequently with soap and water or with alcohol-based cleaners.

- Avoid close contact with others while infected.

- Ask the doctor or pharmacist about antibiotic resistance.

Finally, to further emphasize the seriousness of the problem of antibiotic resistance, the CDC recommends the following *"DO NOTs"* for consumers:

1. *DO NOT* take antibiotics for virus infections such as common colds, most bronchitis, coughs, or the flu. *Do, however, follow the doctor's instructions.*

2. *DO NOT* demand antibiotics from the doctor when told that they are not needed.

3. *DO NOT* skip any doses of prescribed antibiotics night or day and take the full course of prescribed antibiotic therapy until either all medication is gone or the doctor says to stop taking it.

4. *DO NOT* save any antibiotics for the next time one gets sick.

5. *DO NOT* take antibiotics prescribed for someone else.

6. *DO NOT* share antibiotics with someone else.

Antibiotic Resistance Is Everybody's Problem

Inappropriate use of antibiotics in agriculture and aquaculture already have had profound consequences for human health. The use of antibiotics in veterinary medicine to treat and cure animal disease is a legitimate use. However, their use to promote growth in food animals crowded together in unsanitary conditions is considered to be an inappropriate use of antibiotics. In fact, public health authorities have linked low-level antibiotic use in conventionally raised livestock directly to increased numbers of people contracting infections that resist treatment with the same drugs or antibiotics in the same chemical class.

In the meantime, it is imperative that prescribing physicians find ways to curtail over-prescribing of antibiotics to patients whose diagnoses are at odds with the indicated uses of these important drugs and their best professional judgment based on their cumulative experience as a medical practitioner; and that our drug regulating agency refuse to approve for animal use any antibiotic in current use in human medicine or in a chemical class which includes antibiotics in current use in human medicine.

Since no one is immune to becoming a victim of antibiotic resistant bacteria, this is everybody's problem and everyone has a responsibility in slowing the proliferation of these deadly organisms.

Chapter 6: Consumer-Driven Medication-Related Problems

In the previous chapter, we started discussing medication-related problems. We began by discussing adverse drug reactions, medication errors, and antibiotic/microbial resistance, over which health care professionals and the health care system exercise control and have both the greater resources and greater responsibility to prevent these problems from occurring to a much greater extent than the individual consumer. This, however, does not completely absolve consumers of their responsibility for cooperating in every way possible to make it easier for their health care professionals to make their drug therapy experiences safer and more effective.

On the other hand, another group of medication-related problems provide the consumer with greater opportunity and responsibility for protecting themselves. All too often, consumers take both prescription and non-prescription drugs—along with dietary and herbal supplements—callously as if they were mere commodities. Consumers ofttimes have the ability to modify behaviors that can make major improvements in these problem areas or can prevent them altogether. In those cases where consumers cannot remedy these problems alone, their active participation and cooperation are required for their health professionals to be able to reduce or prevent them altogether. These are the medication-related problems we will now discuss in this chapter:

1. *consumer non-compliance*, also referred to as non-adherence, which includes intentional and unintentional consumer failures to follow instructions for taking prescribed and non-prescription drugs, and associated warnings and cautions;

2. *drug misuse, overuse and abuse* of prescription and non-prescription drugs, or the abuse of alcohol, tobacco and illicit drug substances; and

3. *consumer health literacy* and *drug literacy*, which is a person's ability to understand and utilize the health care system, and drug and other health information in beneficial ways.

The purpose of this chapter is to provide consumers with sufficient knowledge and understanding of preventable, consumer-driven problems associated with taking prescription and non-prescription drugs alone or in combination with other substances; e.g., with other drugs, dietary supplements, herbal supplements, alcoholic beverages, certain foods, and certain diseases or health conditions, over and above the preventable medication errors caused by health care providers, care-givers and consumers themselves. After reading this chapter, hopefully, the reader will better appreciate the need for exercising a cautious optimism through more intelligent use of these healing miracles.

Consumers are doubly warned about ignoring instructions from prescribers and pharmacists. The information in this chapter will provide the basis for achieving these objectives by enabling consumers to apply what they learn by more actively participating on their own health care teams with their doctors and other health care providers. They will be empowered to ask more informed questions, make more helpful observations, explore healthier ideas, and demonstrate greater understanding and compliance.

Consumer Non-Compliance and Non-Persistence

Consumer non-compliance or non-adherence to drug dosage regimens is a multi-dimensional problem which adds significantly to individual consumer drug costs, a consumer's overall health care costs, and the public's costs for health care in the United States. In fact, it adds an estimated $100 billion in direct costs to the U.S. health care system annually. It also adds an estimated $1.5 billion in indirect costs from lost wages and some $50 billion in lost productivity annually.

Non-compliance or non-adherence refers to a person's failure to follow verbal, written or printed instructions for taking prescription or over-the-counter (OTC) medication appropriately. Non-compliance with medication instructions has long been a major, pervasive and costly problem in health care. Compliance or adherence with drug regimens is defined as the degree to which consumers take their medications as prescribed by their doctors or other prescribing health care providers.

The World Health Organization prefers terminology which is more suggestive of the active participation of consumers in decisions about their own health care, rather than their being expected to blindly comply with directives from doctors and other health care professionals. The term "adherence" is suggestive of consumers' involvement in their drug therapy decisions which, of course, is felt to enhance the probability that they will more diligently follow therapy instructions "agreed upon" with their prescribers.

As a result, the terminology promoted for use by WHO includes the terms "adherence" and "persistence," rather than "compliance." The term "persistence" refers more directly to consumer continuation or discontinuance of prescribed drug regimens whether for short-term or chronic drug therapy. Nonetheless, while these terms will be used interchangeably in this book, the terms "compliance" and "non-compliance" will be used most often as American consumers are more likely to understand this terminology.

In chronic diseases, the persistence of a person in taking their medication appropriately over the long term is very important. Consequently, the concept of "persistence" is also more frequently used to educate and promote better consumer compliance with drug dosage regimens over a required course of therapy—which may be for many years or for life. Persistence, then, is defined as the extent to which a person follows his or her drug regimens over the intended duration of therapy, whether for days, months or years. This especially applies to the customary long-term or life-time treatments for chronic diseases and conditions like arthritis, diabetes, hypertension, abnormal cholesterol, certain mental disorders, etc. However, it is also applicable to short-term drug therapy as well.

Concepts and terminology, prevalence of, statistics on, factors affecting, and the consequences of non-compliance are discussed in a following section. While under-use and overuse of medication can be due to over-prescribing and under-prescribing, both of which are vital concerns in patient care, they also can simply be the result of patient non-compliance with prescribed drug dosage regimens or, in the case of over-the-counter drugs, simply failure to follow printed label instructions and warnings. In other words, non-adherence—whether intentional or inadvertent—can result in under-medication or over-medication of or serious fatal injury to patients.

Non-compliance leads to treatment failure, adverse drug reactions, emergency room visits, hospitalization or extended hospital stays, severe injury or even death, depending on their existing health status and other circumstances.

Prevalence of Consumer Non-Compliance

As reported by the American Heart Association, some telling drug compliance statistics on U.S. consumers are:

- Some 12 percent of persons don't fill their prescriptions at all;
- Some 12 percent of persons don't take their medications at all, even after filling their prescriptions;
- Almost 29 percent of persons stop taking their medication before taking all of it; and
- About 22 percent of persons take less medication than their prescription label instructs them to take.

For Americans aged 50 and older, the American Association of Retired Persons (AARP) reported the following as the main reasons for not filling their prescriptions based on the results of a 2004 survey:[35]

- Cost of the drug..40 percent
- Side-effects of the drug ....11 percent
- Belief that drug would not help ....11 percent
- Belief that they did not need the drug..........................8 percent

- The drug did not help...6 percent
- Dislike for taking drugs...5 percent
- Their condition improved..4 percent
- Already taking too many prescriptions ....3 percent

Former U.S. Surgeon General C. Everett Koop is quoted as saying that "Drugs don't work if people don't take them." What a simple but most profound statement this is regarding prescriptions!

Another startling revelation comes from the National Council on Patient Information and Education (NCPIE). According to the NCPIE, some 60 percent of prescribed medication is taken incorrectly or not at all. Another author reported that among consumers who understand and actually agree with their treatment, only about 75 percent are likely to take their medications as prescribed.

Other important but disturbing findings about patient non-compliance with prescription drug regimens have been reported. For instance,

1. About a third of all prescriptions are never filled and improper administration is found to occur in more than half of those filled.

2. In patients prescribed high blood pressure and cholesterol lowering medication to be taken together, only one in three was found to be taking both medications for a full six months.

3. Another study of almost a quarter million patients newly prescribed an antidepressant, after not having taken one during the previous six months or longer, found that fewer than 30 percent were still taking their antidepressant for the full six months.

4. In still another study of elderly patients started on statin medication to reduce their risk of coronary heart disease, only about 26 percent were found to continue high levels of use five years later.

5. A 2002 survey of 10,000 patients revealed that reasons for non-adherence are multidimensional. That is, the reasons stated for

non-adherence by the patients surveyed varied considerably as follows:

a. About 24 percent of participants gave forgetfulness as the reason for their non-adherence.

b. Some 20 percent blamed medication side-effects.

c. Some 17 percent blamed high drug costs.

d. Some 14 percent felt they didn't need the drug.

e. Others reported not knowing how to use the drug, being unable to get the prescription filled, picked up or delivered, or various other reasons.

These aggregated findings point to one of the most serious problems in health care involving drug therapy, second only to the problem of medication errors resulting from prescriber and health system failures. Simply pointing up the severity of the problem, of course, does little to get at the roots of the problem. It is important for consumers and health care providers to better understand root causes of this widespread threat to patient health and safety.

Causes of Non-Compliance and Non-Persistence

The following types of behavior can cause problems by consumers or caregivers after prescription and non-prescription drugs leave the pharmacy:

1. Failure to read the labels.

2. Failure to clarify with the prescriber or pharmacist terms or directions they do not understand.

3. Failure to read drug information leaflets accompanying prescriptions.

4. Failure to measure or count out the medication properly.

5. Failure to take the right amount of medication.

6. Saving medication to make it last longer by reducing the dose taken.

7. Sharing medication with other family members or friends intentionally or inadvertently for legitimate medical reasons or for illicit use.

8. Failure to take medication according to the schedule prescribed.

9. Removal of drug units from original prescription container to an unsafe container.

10. Mixing different tablet and capsule medications in the same vial or other container.

11. Defacing or removing the prescription label and taking or administering medication from memory without consulting the prescriber or pharmacist or seeking to obtain a replacement label.

12. Improper storage of prescription containers in humid, hot areas, and without regard to other safety requirements, including leaving tops or closures off containers.

Safe medication use by consumers is a consequence of appropriate education by health care professionals and consumers' attitude, knowledge, health status and literacy skills. Accordingly, the following are some of the most important issues affecting how safe and satisfactory a consumer's drug-taking experiences will be:

1. Attitude toward taking medications and toward complying with instructions of the doctor, nurse or pharmacist;

2. Relationship and communication with the doctor;

3. relationship and communication with the pharmacist;

4. Understanding and commitment to complying with medication instructions, cautions and warnings given by the doctor, nurse and pharmacist;

5. Attitude toward handling and safeguarding of medications around children;

6. Understanding of how differences in age, sex and ethnicity can affect their medication use and treatment outcomes;

7. Attitude toward prohibitions on sharing medications with family members, friends and others;

8. Knowledge and attitude toward storage and handling of one's medications;

9. Knowledge and understanding of basic differences in the most commonly used dosage forms and how these differences affect medication dosing and appropriate storage and handling requirements.

Costs and Consequences of Non-Compliance

Over the past two decades, the consequences and costs of consumer non-adherence with medication have been documented quite well through various studies reported in the literature. Based on one such study, the rate of non-adherence to prescribed medication is estimated to average about 50 percent.[36]

The costs and consequences of non-compliance to individual consumers and society are heavy both in dollars and cents and in human terms. For instance, it is estimated that non-adherence adds to the cost of health care in the United States by adding direct costs to the tune of some $100 billion annually; and indirect costs of some $1.5 billion in lost wages, and some $50 billion in lost productivity annually.

In human terms, some 125,000 deaths result annually from patients' failure to adhere to prescribed medication. Additionally, about 10 percent of hospital admissions and some 23 percent of nursing home admissions are the result of consumer failure to adhere to prescribed medication.

Prescription Drug and Other Substance Abuse

After taking some drugs for a while, the body requires increasingly larger doses in order to get the same relief. This effect is referred to as *drug tolerance*. Drug tolerance occurs when the accustomed dose of a drug no longer has the same therapeutic effect or no longer gives the same response. Therefore, to receive the same response from the drug, one must take increasingly larger doses.

Drug dependence occurs when a patient or consumer has developed either a physical (or physiological) craving or psychological need for a drug substance. A *drug addiction* is a physical or physiological craving where the lack of a drug—legal or illicit—causes severe withdrawal effects in the person.

Widespread adult practices of reaching for a drug for every discomfort, taking a drink to smooth the edges, and reaching for a smoke to soothe our anxieties, set a poor example for our youth and help create a culture of casual and recreational pill popping among our youth. The result of these adult examples, coupled with the widespread glorification of similar behaviors in the entertainment world, further encourages our youth along the road to self-destructive behavior such as the intentional misuse and abuse of prescription and illicit drugs.

Types of Prescription Drugs Commonly Abused

Results of a recent national survey by the National Institute on Drug Abuse (NIDA) showed some 19.7 million Americans 12 years or older were users of illicit drugs. Further, non-medical, recreational use of prescription drugs by young adults from ages 18 to 25 years of age increased from 5.4 percent in 2002 to 6.3 percent in 2005, primarily due to an increased misappropriation and use of prescription pain relievers.

Most consumers take prescription drugs responsibly, and addiction to prescription drugs is rare, according to the NIDA. However, a large number of individuals report using prescription drugs for non-medical reasons at least once a year. The most frequently abused types of prescriptions drugs are:

1.   Opioids and morphine derivatives;
2.   Central nervous system (CNS) depressants; and
3.   Central nervous system stimulants.

Opioids and morphine derivatives are medications used for pain relief and include codeine, fentanyl, morphine, opium, hydrocodone, hydromorphone, meperidine, oxycodone and propoxyphene. Popular brand names in this group are Darvon, Darvocet, Dilaudid, Demerol, Dilaudid, Empirin with Codeine, Fiorinal with Codeine, Lortab, Lorcet, Roxanol, Oxycontin, Percocet, Percodan and Tylox. Depending on the product, these may be swallowed, chewed, crushed, smoked, snorted or injected.

The most abused among the *stimulant prescription drugs* are the amphetamines, cocaine, methamphetamine and methylphenidate. Popular brand names are Biphetamine, Dexedrine, Desoxyn and Ritalin. These are usually injected, smoked, snorted or swallowed.

CNS depressants most frequently abused include the barbiturates and benzodiazepines. Among the most popular sedative barbiturates are Amytal, Nembutal, generic phenobarbital, and Seconal. Ativan, Halcion, Librium, Valium and Xanax are the most popular benzodiazepines abused. Most of these are swallowed but some may be injected and snorted.

Prevalence of Prescription and Substance Abuse

Non-medical use of prescription drugs among young adults in the 18 to 25 year age group increased from 5.4 percent in 2002 to 6.3 percent in 2005, primarily due to an increase in the use of prescription pain relievers.

Based on results of the 2005 National Survey on Drug Use and Health,[37] an estimated 19.7 million American 12 years or older were users of illicit drugs. Prescription psychotherapeutic drugs used for non-medical reasons were included along with marijuana/hashish, the most commonly used illicit drug (14.6 million current users); cocaine (including crack cocaine), used by 2.4 million abusers; and hallucinogens including Ecstasy, used by 1.1 million persons. Prescription type drugs abused or used for non-medical reasons by some 6.4 million individuals included pain relievers (4.7 million users), tranquilizers (1.8 million users), stimulants, including methamphetamine (1.1 million users), and sedatives (272,000 users).

The use of illicit drugs is estimated to cost the United States some $110 billion annually. Like alcohol, illicit drug use is more prevalent among the unemployed than among the fully employed. Many addicts are also heavy drinkers, although only a minority of alcoholics are drug abusers. Specific illicit drugs with the highest levels of abuse in the 2005 report were marijuana (4.1 million users), cocaine (1.5 million) and pain killers (1.5 million).

Of the estimated 22.2 million persons over 12 years of age classified as having substance dependence in 2005, 3.3 million were classified as abusing both alcohol and illicit drugs, 3.6 million were classified as being dependent on illicit drugs but not alcohol, and 15.4 million were abusers of alcohol but not illicit drugs.

Alcohol Use and Abuse

Alcohol abuse is estimated to cost some $166 billion annually. The $18 billion spent on alcohol and drug treatment in 2005 represented 1.3 percent of all health care spending.

Some 51.8 percent (126 million) of the 12 years and older group surveyed were current drinkers of alcohol and some 22.7 percent or some 55 million persons participated in binge drinking, while some 6.6 percent or about 16 million reported heavy drinking. Within the 18 to 25 year group of young adults, the rates for binge drinking (41.9 percent) and heavy drinking (15.3 percent) remained relatively unchanged from the previous years 2002, 2003 and 2004. However, there were small declines in use and binge drinking in the 12 to 17 year group of youngsters in 2005 compared to 2004 although there was no significant change in the prevalence of heavy drinking in this age group.

Of the 10.8 million drinkers in the 12 to 20 year group (28.2 percent of this age group), almost 7.2 million (18.8 percent) were binge drinkers and 2.3 million (6 percent) were heavy drinkers.

Based on survey data for racial/ethnic groups, alcohol use was highest among whites at 32.3 percent, followed by Hispanics at 25.9 percent; those reporting 2 or more races, 24 percent; American Indians or Alaska Natives, 21.7 percent; African-Americans,19 percent; Asians, 15.5 percent; and Native Hawaiians or Other Pacific Islanders, 12 percent.

Tobacco Addiction and Smoking

Smoking is estimated to cost some $157 billion annually, including some $75 billion in direct medical expenses, related costs and absenteeism and lost productivity from ill patients. Some feel that the low-tax/untaxed underground cigarette sales via the Internet and Indian reservations makes it unlikely that sales and usage have dropped much over the past decade, in spite of any government statistics to the contrary. Further, the major cigarette producers are again in the limelight for alleged practices of promoting low tar products while slowly but steadily increasing the nicotine content.

An estimated 71.5 million Americans 12 years or older were smokers, representing 29.4 percent of this age group. This number was distributed among 60.5 million cigarette smokers (24.9 percent of the group); 13.6 million cigar smokers (5.6 percent of the group); and 2.2 million pipe smokers (0.9 percent of the group). Also within this age group, current rates of use of cigarettes, smokeless tobacco, cigars and pipe tobacco were about the same from 2004 to 2005.

Further details of tobacco use trends, first-time use, consequences and costs, youth prevention measures and treatment related issues are covered later in this book also.

In a recent report, the CDC reported that a significant proportion of suicides during 2004 involved substances of abuse, based on information received from 13 states participating in the Violent Death Reporting System. The substances most frequently identified among the suicide victims were alcohol (33.3 percent), prescription opiate analgesics (16.4 percent), cocaine (9.4 percent), marijuana (7.7 percent), and amphetamines (3.9 percent). Unfortunately, the various states did not uniformly test for all of these drugs routinely for the data supplied to CDC for the 7,277 suicide cases reported to the system. Yet, according to CDC, the overall data suggested a strong role of substance abuse in suicide.[38]

Health Literacy Affects Drug Use

The importance of health literacy for health care consumers is indicated by the inclusion of it as an objective in the nation's Healthy People 2010 plan.[39] As such, health literacy has been identified as an important component of health communications, medical and pharmaceutical product safety, and oral health.

Among the most important of these as relating to consumer responsibility and the ability of consumers to fend better for themselves is adult literacy. In this case, this term is an umbrella term for basic literacy and health literacy, which includes drug literacy. Due to the profound impact of consumer literacy on individual safety and consumers' ability to benefit from effective and affordable medication use, health and drug literacy is discussed in this chapter.

Imagine for a moment, just how few consumers are sufficiently aware of the meaning and significance of health literacy in their lives and the lives of their family members. In fact, health literacy is so very fundamental to communications between consumers and their doctors and other health care providers and between health care professionals and their patients. It is surprising that the subject has been relegated to such low priority status as a health care issue in the United States when the payoff from higher promotion and financing can have such significant payoffs in improved health and financial savings to our health care system.

Health literacy can be defined in many ways but the objective is for everyone to understand how to obtain and use whatever information and other resources are needed to protect their own health and that of their loved ones. This book is focused on improving the *"drug literacy"* of the average consumer so that the maximum benefits can be achieved with minimal risk by consumers who are involved in drug therapy as an integral component of their health care.

What Health and Drug Literacy Is All About

Health literacy is the ability of consumers to read and understand health-related information and to use it effectively. According to the nation's prestigious Institute of Medicine (IOM):

> *Health literacy is the degree to which individuals can obtain, process and understand the basic health information and services they need to make appropriate health decisions.*[40]

The IOM reported that nearly half of all American adults have difficulty understanding and acting upon health information. Further, consumers encounter complex health information frequently, including books, magazines and other periodicals, insurance information and forms, applications for medical assistance, health reports, provider instructions, prescription and over-the-counter medication labels, vitamin and mineral supplement labels, etc.

Even consumers with above average literacy levels and well educated people ofttimes encounter difficulty obtaining, understanding and using "complex" information in the above categories of information. The importance of health literacy extends far beyond the individual as it is dependent upon the expectations, preferences and skills of our doctors (physicians, dentists, podiatrists, optometrists, etc.) and other prescribers, including nurse practitioners, pharmacists and psychologists (in some states); the media; and others involved in collecting, analyzing, reporting and otherwise communicating health information.

In areas related to drug therapy, health literacy includes the ability to understand printed instructions on prescription drug bottles and vials, doctor's directions, drug identification information, drug expiration dates, drug information handouts, counting and measuring doses and devices, scheduling, refilling prescriptions, appointment slips, medical education brochures, and consent forms, and the ability to negotiate complex health care systems. However, health literacy involves much more than simply the ability to read. It involves the command of a complex group of reading, listening, analytical, and decision-making skills, as well as the ability to apply these skills to maintaining or solving problems related to health.

Drug literacy, then, is one component of the broader term "health literacy" and, while considering the overall problem of health literacy in health care in America, our discussions will pinpoint those issues and problems related to safe and rational drug therapy for the average person using prescription and/or non-prescription drugs.

One of the biggest fallacies about literacy is that one's level of education is the absolute determinant of health literacy. Admittedly, education is a major factor in health literacy. However, authorities agree that health literacy varies by context and setting and that it is not necessarily related to a person's years of education or their general reading ability. Many people who function well at home and work encounter difficulty in navigating health information and the health care system because of their marginal or inadequate literacy in health care.

Several major factors affect consumers' health literacy and, consequently, drug literacy. These include age, education, culture, language, social services and health services.

Reading ability is a vital component of all health literacy skills and marginal reading skills are defined as reading abilities below 8th grade level. Low level reading abilities are effective barriers to understanding and using health information. Complex text is quite common in health information and consequently in drug related information.

The scope of the problem is indicated by the following statistics from a national survey:

1.   One in five adults, or 20 percent, reads at or below the fifth grade level;

2.   Almost two of five or 40 percent of Americans age 65 or older read at or below the fifth grade level;

3.   The average reading level for adult Americans is between the eighth and ninth grade;

4.   Most health materials are written at the 10th grade level or above (three-fourths);

5.   Nearly half of all American adults have difficulty understanding and acting upon health information;

6.   The language in informed consent forms and consumer privacy notices are often not readable by some 40 million Americans due to their complexity.

According to the 2003 National Adult Literacy Survey, the consumer's ability to understand and use health related information effectively has important consequences for overall health care costs and personal and family safety.[41] Of those persons in the survey who were found to be at or below the fifth grade level of literacy:

1.   Some 60 percent were over 60 years of age;

2. Some 25 percent were immigrants;

3. Some 77 percent had less than an 8th grade education;

4. Some 7 percent had visual problems; and

5. Some 3 percent were dyslexic.

The ethnic breakdown of the adult population in the group surveyed in this study was: white, 15 percent; Native American, 26 percent; Asian/Pacific Islander, 35 percent; African-American, 41 percent; and Hispanic (all groups), 52 percent. Due to the majority status of native born white Americans, they actually constituted the largest number of those at or below the fifth grade literacy level.

Consumer literacy is affected by many factors. Health literacy may be impacted by language and cultural barriers, mental/cognitive impairment, disease and other medical conditions, the effects of medications they are taking, etc. After all, health literacy is about more than reading, writing and mathematical skills. Other important skills and abilities involve speaking, listening, self-advocacy and the ability to use varied information and use it to come up with solutions to health care problems of various levels of complexity.

For instance, patients need to be able to ask questions about their drug therapy during their doctor visits, as well as at the pharmacy when they leave their prescriptions to be filled and when they pick up their prescriptions once they are filled. They need to be able to read and understand the instructions on the printed prescription label, the accompanying cautionary and warning labels, as well as the drug information pamphlets given to them by the pharmacist with their prescriptions. They need to know the specific reasons for the medications they are taking, how they are to take them properly and safely, how much medication they are getting or should be getting for each prescription, how long they should take them, the number of refills authorized, if any, and any unusual circumstances or potentially worrisome or serious side-effects they may encounter while taking the mediation. Patients need to know what to do if they encounter adverse effects from the medication they are taking and when they should call their doctor or pharmacist with problems they encounter while on the medication.

Consumers need to be able to read and understand the active ingredients, strengths, directions for use, length of time to use, and any cautions and warnings about the non-prescription drugs they purchase on their own. They will then be able to compare different non-prescription drug products by comparing the active ingredients, dosage strengths, directions for use, length of time to use, and any cautions and warnings on the labels. This will enable them to prevent taking double or triple doses of a particular active drug when they take two or more of the over-the-counter products they buy to relieve pain, headaches, coughs and colds. Consumers certainly need to know when they should stop self-medicating and when to seek professional medical attention. The consequences of not being able to read and understand the information on product labels can be grave.

Drug Literacy Costs and Consequences

Let us consider some of the consequences of low literacy on health in these United States. First, it is estimated that the annual costs to the U.S. economy for low health literacy are between $5 billion and $9 billion. It is already tragic that about half of all patients make medication errors, but even more tragic to see less literate consumers being much more likely to make medication errors during their treatment. Studies have further shown that less literate consumers are much more likely to make medication errors than consumers with higher literacy rates. Consumers with marginal literacy skills are more apt to make more serious medication errors and other treatment errors, and are less likely to be able to comply with medication and other treatment directions. They are also much less likely to seek preventive care partly because they are less able to navigate the health care system than their more literate counterparts.

Examples of the types of medication-related problems encountered and errors made by patients with marginal or low literacy skills are:

1. Approximately 23 percent don't know how many refills they have on their prescription;

2. About one-third, or 33 percent, make errors in taking or administering their medications. The following are some of the reported examples of such problems:

    a. One mother thought that instructions for a child to "take 3 times daily" meant to give her child one-third of the contents of a bottle of the medicine each time.

    b. One man applied one skin patch every 6 hours until he had multiple patches on at the same time, instead of removing the old patch each time he placed a new patch to his skin.

    c. A 66-year old man, took two dosages of a blood thinner daily instead of alternating the dosages every other day, as was intended, with no followup for 2 weeks, and with dire consequences which landed him in a hospital emergency room.

Patients with limited English proficiency (LEP) were found to be more likely to experience adverse events in U.S. hospitals and to suffer more severe harm in so doing, according to one pilot report.[42] Prominent among causes were questionable advice or interpretation of the advice or instructions at a rate more than 2.5 times that for English-speaking patients (11 percent versus 4 percent). The results were similar, although a bit higher, for adverse events attributed to questionable assessment of patient needs (15 percent versus 6 percent). The researchers also found that 52 percent of 251 adverse events involving LEP patients at six hospitals over a seven-month period related to communication problems. This compared to 36 percent for English-speaking patients for communication-related adverse events.

Almost a third of adults over the age of 65 years are estimated to have very poor health literacy skills; 41 percent of Hispanic adults are estimated to have poor health literacy skills; and some 24 percent of African-American adults are reported to have poor health literacy skills.

While low health literacy affects all of us, it affects some more than others.[43] For instance, two out of every five adult Americans encounter difficulty in obtaining, processing, and understanding basic information and services needed to make appropriate decisions for their health. Consumer health literacy is a major public health problem in the United States. While the federal government, some national medical organizations, and a few major brand-name pharmaceutical firms are sponsoring programs targeting this important consumer literacy problem, either directly or indirectly, current needs far outstrip current and projected resources being allocated to this major health problem.

The pharmaceutical industry must show a much larger commitment to this cause for more than purely altruistic reasons, but also because consumer compliance and persistence is good for the industry. The U.S. government should be willing to commit much greater resources to this effort because the health and financial returns on such an investment would help transform the character of our national health care system into one of higher quality, improved treatment outcomes, and lower per capita costs, while greatly increasing the nation's overall productivity and, therefore, substantially improving the American economy.

Consumers themselves, however, must also assume greater responsibility in seeking out personal, professional and printed resources through local and national government, civic, religious, and professional organizations to enable them to improve their own literacy levels. This will involve, among other actions, actively improving their proficiency in the English language and becoming more proactive about participating in community health programs as much as possible and reading and listening to as many educational and health-related communications as possible in newspapers, magazines, on the radio and on television in preference to many of the currently preferred but less helpful communications and programs they presently use.

SECTION III: Industry, Government, and Consumer Alternatives

Chapter 7: Industrial Profits at Consumer Expense

It would be counterproductive even to try to ignore the awesome benefits of prescription drug discovery and development to our world civilization and its enormous, incalculable benefits in saving and improving countless millions of lives, even in the face of well-documented cases of anti-consumer, anti-competitive behavior on the part of members of America's drug industry machine called Big Pharma. Yet the alarm must be sounded to alert our citizens and responsible elected local and national officials of specific industry-related and government-related issues which seriously impede consumer ready access to safer, more affordable drug therapy in these United States.

Many venture to claim that the medical profession is either in league with Big Pharma or so overwhelmed by it in creating new disease conditions for existing drugs to maintain and build market share to the detriment of the earlier availability of cheaper, equally effective, generic drug substitutes. Still others claim that Big Pharma's heavy and expensive advertising of brand-name prescription medication to prescribing physicians unduly influences prescribing habits, and that their heavy direct-to-consumer advertising increases consumer demand for the more expensive brand-name, single-source prescription drug products manufactured by Big Pharma members. Unarguably this is financially healthy for the drug manufacturing giants.

A frequent question is whether the increasing per capita level of consumer drug use is as healthy for the individual and overall public health as to justify the risks inherent in it and the overall costs being paid for it. This is a fair and highly pertinent question which information presented in this chapter should enable motivated drug consumers and committed health care professionals to better answer for themselves.

The balance between cost *versus* benefit all too often has been tipped disproportionately in favor of the pharmaceutical industry behemoth at the expense of consumer health, safety, and access to more affordable prescription drugs, in large part due to ineffectual enforcement of post-market safety surveillance studies by the FDA. This is also, at least in part, a consequence of the forced reliance of the FDA on major pharmaceutical industry funding, as official government policy, to perform its mandated new drug regulatory functions, coupled with a lack of clear legislated authority to fulfil its drug safety mandate with the independence this agency deserves and requires.

Beyond these, an ever increasing number of anti-competitive, anti-consumer strategies continue to emerge in the prescription drug marketplace. More recently, pro-consumer, pro-competition actions taken by the once powerful Federal Trade Commission (FTC) to prevent brand-name and generic pharmaceutical firms from conniving to delay the marketing of new generic drugs have been overturned by our nation's top judiciary, the Supreme Court. Consumers, as well as health care practitioners, need to be alert to these continually evolving challenges to consumer safety, health and welfare. The thrust of this chapter is to identify and discuss some of these major challenges and provide some background information to serve as a basis for consumer empowerment.

Such a case represents one of the colossal failures of government and industry in fulfilling their duties as guardians of the public health. That is why it is incumbent upon consumers, consumer advocacy groups, health care professional groups, and our duly elected officials to better educate themselves and the general public about the issues and become more intensely proactive in demanding and taking appropriate action to ensure that matters of consumer drug safety, access and affordability are at the top of America's public health agenda.

Paradoxically, at the other end of the drug use spectrum is an estimated 50 percent rate of non-compliance or non-adherence to medication completion as instructed, thereby adversely impacting intended treatment outcomes from prescribed medications. This also hinders legitimate efforts to better quantify actual consumer drug usage, since prescription drug purchases clearly do not necessarily equate to drug use.

Drug Prices, Profits and Prevalence of Use

A known fact, however, is that the U.S. prescription drug manufacturing industry is among the most profitable of the nation's major industries and has been for years. Ask the average consumer who purchases prescription medications for themselves or family members. Ask them what their single most important question about prescriptions is and discover that the first question usually asked by consumers will be, "Why are drug prices so high?" And, right or not, consumers equate "high prices" to high profits. While retail pharmacies in general enjoy healthy profits, this discussion will show that the manufacturer's price is by far the major component of a prescription's final cost to the patient rather than the pharmacy's markup to cover expenses and a reasonable profit.

Concerns about the "high cost" of drugs are fair and important because they bear directly on the issue of consumer access to affordable medication for all, but this is especially troublesome for the most vulnerable segments of our population. Yet, the answers to these concerns are complex and raise additional and sometimes troubling questions about prescription drug safety, cost versus benefits, and ultimate value. Simplistic answers fall short of shedding light on these important issues.

At the same time, it is instructive to know some of the most important reasons for higher brand-name prescription prices and the high level of profitability for pharmaceutical manufacturers in general. These include the following:

1. *Heavy advertising and promotion of brand-name drugs* to both physicians and consumers, the costs for which are recouped in the prices charged the pharmacy and ultimately the consumer or other payer for these prescription drugs.

2. *Drug prices set sufficiently high* to enable the innovator drug companies to recoup heavy research and development, capital

costs for manufacturing the new drug, and initial product launch costs as soon as possible, to provide reasonable returns for investors, and to provide cushions for future discovery of new drug therapies and potential product liability cases.

3. *Legal monopolies conferred* on new drug manufacturers *through patent protection* for periods up to 20 years as incentives for these companies to make such heavy investments in new drug discovery and development.

4. *Patent defenses* launched by the major pharmaceutical manufacturers, including such controversial practices as promoting authorized generic versions after patent expiration, bribing generic manufacturers into delaying introduction of their cheaper first-generic versions of previously patented brand-name drugs, and filing last minute and frivolous citizens petitions with the FDA in order to delay or prevent bonafide generic competition for as long as possible.

The Most Frequently Prescribed Drugs

While there are thousands of pharmaceutical manufacturers and thousands of prescription and non-prescription drug products, the bulk of brand-name and generic drugs are provided by a small number of producers. Data presented in this section will focus on the top 10 prescription drugs and top 10 manufacturers of both branded and generic prescriptions used in the United States during the most recent year for which data are available at the time of writing this book. The data presented include the brand name of the drug, the generic name, the manufacturer of the brand-name innovator drug, and either the number of prescriptions filled with that drug at retail or the drug's dollar sales volume in the United States for the designated years.

Based on the number of prescriptions dispensed in the United States, the following 10 drugs topped the list during 2006:[44]

1. Lipitor (atorvastatin), Pfizer .... 74.0 million prescriptions

2. Hydrocodone/acetaminophen, multiple sources .... 70.0 million prescriptions

3. Toprol XL (metoprolol), AstraZeneca . 41.4 million prescriptions

4. Norvasc (amlodipine), Pfizer .... 40.0 million prescriptions

5. Amoxicillin, generic multi-source drug 34.2 million prescriptions

6. Hydrocodone/acetaminophen pain drug combinations (in addition to the above), multiple sources .... 33.3 million prescriptions

7. Synthroid (levothyroxine), Abbott ...... 30.9 million prescriptions

8. Nexium (esomeprazole), AstraZeneca.. 30.4 million prescriptions

9. Lexapro (escitalopram), AstraZeneca... 30.2 million prescriptions

10. Albuterol, multiple sources .... 29.6 million prescriptions

However, *the top 10 individual prescription drugs by dollar sales* in the United States during 2006 were major brand-name drugs and included three biopharmaceuticals.[45] These were:

1. Lipitor (atorvastatin), Pfizer .... $8.6 billion

2. Nexium (esomeprazole), AstraZeneca .... 5.1 billion

3. Advair Discus, GlaxoSmithKline .... 3.9 billion

4. Aranesp (darbepoetin alfa), Amgen .... 3.9 billion

5. Prevacid (lansoprazole), TAP .... 3.5 billion

6. Epogen (epoetin alfa), Amgen .... 3.2 billion

7. Zocor (simvastatin), Merck .... 3.1 billion

8. Enbrel (etanercept), Amgen .... 3.0 billion

9. Seroquel (quetiapine), AstraZeneca .... 3.0 billion

10. Singulair (montelukast), Merck .... 3.0 billion

The Most Frequently Prescribed Drug Classes

The following were the top 10 therapeutic classes of drugs prescribed in the United States based on sales dollars paid to brand-name drug manufacturers during the 12-month period ending December 2006:[46]

1. Cholesterol/triglyceride reducers .... $21.6 billion
2. Protein pump inhibitors/anti-ulcer drugs .... 13.6 billion
3. Antidepressants .... 13.5 billion
4. Antipsychotic drugs (Other) .... 11.5 billion
5. Erythropoietin cancer drugs .... 10.0 billion
6. Epilepsy/seizure disorder drugs .... 8.9 billion
7. Antineoplastic monoclonal antibodies .... 5.8 billion
8. Angiotensin II antagonists .... 5.7 billion
9. Calcium blockers .... 4.7 billion
10. Oral diabetes drugs .... 4.1 billion

The top 10 therapeutic classes of prescription drugs *based on the number of prescriptions dispensed* in the United States in 2006 were the following:[47]

1. Antidepressants .... 227.3 million
2. Lipid regulators .... 203.0 million
3. Codeine and combination drugs .... 177.1 million
4. ACE inhibitors for high blood pressure, etc. .... 154.1 million
5. Beta blockers for high blood pressure .... 130.4 million
6. Proton pump inhibitors .... 101.6 million
7. Thyroid hormone (synthetic) .... 97.6 million
8. Epilepsy/Seizure disorder medications .... 94.9 million
9. Calcium blockers for high blood pressure .... 86.9 million
10. Benzodiazepines .... 80.1 million

Raising Awareness of the Problem and Related Issues

Because of this, it is paramount that consumers, payers and care-takers, as well as their health care providers, raise their awareness of the flip side of the pill-shaped coin on the important issues of patient safety, treatment outcomes and drug costs. Indeed, increasing this awareness of these issues is the one of the most important objectives of this book.

For consumers, this means that they must actively better inform themselves, to the extent possible and by every possible means, of drug industry practices which adversely impact the ready availability of less costly generic drugs and the prices consumers must pay for their prescriptions. They must know of the unlawful behaviors engaged in predominantly by the brand-name segment of the prescription drug industry, especially the intentional forays by drug firms into gambling that their projected sales and profits from such practices will more than adequately compensate them for any later legal judgments they may face. To attain and maintain any reasonable momentum in this direction, it is vital that consumers question the controversial practices of the drug manufacturing industry even those which may be technically lawful, when these are deemed to be counter to the spirit of the law, and certainly those which are patently unlawful, negligent or injurious to the public health and welfare.

Public Duty and Accountability of Industry and Pols

The time has come to hold both industry and our national political leaders accountable for funding deficits, misplaced health system priorities, and negligence in protecting consumers' safety and easier access to more affordable prescription (and non-prescription) medication. The nation's leaders and legislators must elevate consumer health as an overriding principle of public health policy to ensure the long-term viability and security of the nation. A more drug literate, more health literate, more involved electorate has the right, responsibility and the power to ensure the existence of such a common sense policy by holding our political leaders accountable for it.

Pharmacists understand and are usually tolerant of consumer anger directed toward them for the high cash prices or high co-pays for their prescriptions. However, the public needs to better understand the forces and factors beyond the pharmacist's control which play the greater roles than the pharmacist or pharmacy in determining the final prescription prices they pay. Pharmacists and prescribers have a duty to inform and better educate their clientele on industry's influences on prescription costs and safety.

The organization representing the major U.S. pharmaceutical producers is the Pharmaceutical Research and Manufacturers of America (PhRMA). Its membership is comprised of about 200 companies and wields almost incomprehensible power over the lives and livelihood of consumers in the United States and abroad.

The pharmaceutical manufacturing industry, as any other business, must be profitable and provide just returns on corporate investments. Indeed, it must be profitable enough to take substantial financial risks by investing heavily in new drug research and development activities with uncertain returns. At the same time, the very fact that medications are far more than mere commodities and that the public accords manufacturers such a high level of trust through their government licensing and regulatory processes, they accordingly owe a high level of responsibility to the consuming public.

The major brand drug manufacturers justify their drug prices and their lucrative profit margins on the basis of their lengthy and expensive research and development costs invested over periods of 10 to 15 years to discover, test, develop, gain FDA approval to market their new drugs, build new production facilities, and promote and market them to prescribers and consumers. Admittedly, the risks these manufacturers take in pursuing new drug discovery and their overall contributions to consumer health and well-being are substantial.

According to PhRMA, the costs to discover and bring a basic new drug agent to market average more than $800,000.000.[48] By the time their products make it to the market, a portion of their customary 20-year patent protection time has lapsed while research and development activities were underway to prove the safety and effectiveness of the emerging new drug from its thousands of earlier fellow lab candidates. The time remaining for patent protection, before less expensive generic versions of their drug can be sold by competitors, requires them to attempt to recoup their costs with higher prices before their monopolies (patent protections) expire, they argue. The law does provide a mechanism for the FDA to compensate these manufacturers with verified patent time lost up to a maximum of three additional years.

On the other hand, it is argued by consumers and consumer advocates that the public trust has been violated and these producers have been grossly unfair at the very least and criminally negligent, at worst, for failing to fulfill their responsibilities to the public in a variety of ways. Accordingly, consumers and their advocates protest that they are injured too frequently and unjustly charged exorbitant prices for the life-saving and life-enhancing drugs they must depend on for survival. The pharmaceutical industry argues conversely that they fulfil their obligations to the public by investing heavily in research and development, and deserve the remuneration they receive for the miracle products they discover, develop, and market.

Several major drug industry practices which affect consumer access to safe and affordable prescription medication are controversial and/or illegal. This includes practices such as heavy promotion of drugs while failing to provide unbiased or complete information on risks associated with their prescription products to both prescribers and patients. These and similar practices have tended to undercut consumer confidence and trust in the pharmaceutical industry.

In this chapter, the reader will find out about certain pervasive and even dangerous consumer protection lapses or brand defenses of the major brand-name prescription drug manufacturing firms, as well as government deficiencies which allow these to transpire. Others like counterfeit drugs are a common enemy to the consumer, the manufacturer and the drug regulatory agency, the FDA. Among the most important of such issues, discussed in this chapter, include the following:

1. *Seeking FDA approval of additional or new indications* (medical uses) for an existing brand-name prescription drug to extend the life of its patent.

2. *Drug patent manipulation and litigation* and other manufacturer strategies used to delay the availability of competitive and affordable generics.

3. *Failure to complete post-market safety studies* by pharmaceutical manufacturers as required and as agreed upon.

4. *Promotion of "off-label" or unapproved uses* by pharmaceutical manufacturers of their FDA-approved brand-name prescription drugs.

5. *Filing of last-minute citizens petitions* to delay "first generics" marketing.

6. *Other brand defenses* perpetrated by the major drug manufacturers to stifle the timely availability of less expensive generic versions of their costlier single-source brand-name prescription products.

7. *Direct-to-consumer advertising* of brand-name prescription drugs.

8. *Other actions which adversely affect the accessibility, affordability and safety of the nation's drug supply*, such as counterfeit drugs and stolen drugs entering the pharmaceutical supply chain.

The Generic Prescription Drug Solution

The prices for generic prescription drugs can represent consumer savings of 30 to 80 percent, compared to the prices for the single-source brand-name drugs they copy. Recently expired and soon to expire drug patents in the United States comprise a long list with the awesome potential of saving millions of dollars for consumers and taxpayers. In fact, the value of major drugs whose patents recently expired or are expiring through 2009 is astounding, as illustrated in Table III below.[49]

YEAR	Major Brand-Name Prescription Drugs Included	SALES**
Table VI. PROJECTED ANNUAL SALES OF RECENTLY EXPIRED AND EXPIRING MAJOR DRUG PATENTS THROUGH 2009*		
2005	Amaryl, Biaxin, Duragesic, Mobic, Oxycontin, Paxil CR, Zithromax	$10.0 billion
2006	Allegra, Maxaquin, Pravachol, Proscar, Provigil, Retonavir, Wellbutrin XL, Zocor, Zofran and Zoloft	$11.3 billion
2007	Actiq, Clarinex, Coreg, Imitrex, Geodon, Lotrel, Mavik, Meridian, Norvasc, Tequin, Zyrtec	$11.1 billion
2008	Advair, Casodex, Depakote, Dovonex, Effexor XR, Fosamax, Kytril, Risperdal, Serevent, Trusopt, Zerit, Zymar	$10.4 billion
2009	Avandia, Lamictal, Lexapro, Prevacid, Topamax	$9.9 billion

Source: Express Scripts 2004 Drug Trend Report via U. S. Pharmacist, June 2006. *Does not include all drug patent expirations during the indicated year; only major expiring drug patents are listed. **Includes sales value for all brand-name drug patents expiring during the year, including other drugs not listed.

After a pharmaceutical manufacturer's drug patent expires, any pharmaceutical manufacturer can produce and sell generic copies of the brand-name product after gaining FDA approval. Consequently, generic prescription drugs are referred to as multi-source products since any number of drug firms can produce and market them. On the other hand, brand-name products are single-source products which command higher prices due to the legal monopoly conferred by their patent protection and their name recognition after patent expiration.

To gain approval of the FDA to market a generic drug, the active ingredient(s) in the generic product must be chemically identical to the active ingredient(s) in the branded drug product. Generics must meet the same strenuous manufacturing and quality control standards as the brands they replace in order to gain FDA's approval. Because the efficacy and safety of the drug substance in a generic has been thoroughly documented in the initial patent holder's NDA and subsequently through widespread medical use during the term of the patent holder's extended monopoly, generic drug producers are not required to rerun clinical and safety studies—except for bioavailability studies—before gaining FDA's approval. Bioavailability is a measure of "the rate and extent to which the active ingredient or active moiety is absorbed from a drug product and becomes available at the site of action" in the body.[50]

The vehicle for generic drug producers to seek FDA approval is the abbreviated NDA (ANDA), which includes only a fraction of the voluminous information required for FDA approval of the initial NDA for a new drug. Because these generic drug producers incur much less expense in bringing their products to market, and in marketing and promotion of their products, the prices for generic drugs can be 30 to 80 percent lower than the prices charged for the brand-name drugs.

This is precisely what was intended when enabling generic drug legislation was first enacted as the Hatch-Waxman Act of 1984. The goal of the Act was to provide multi-source generic drug manufacturers with sufficient incentive to rush their less expensive versions of previously patented drugs to the market. Drug consumers and tax payers would be the principal beneficiaries of the resulting enormous savings.

Generics Delayed Amid Big Pharma Brand Defenses

Businesses must be free to responsibly seek every legally permissible opportunity to profit from their heavy investments in technology, research and capital improvements. However, where the public health is involved, certain actions of industrial pharmaceutical giants will naturally be perceived as being profit-seeking at the expense of the higher public interest of protecting the health and welfare of consumers. Because the courts have given the green light to some of their actions viewed by many as being anti-consumer, both generic firms and firms defending their major brands have become increasingly complicit in recent years in delaying the introduction of the more consumer-friendly new generics.

The following controversial actions taken by the major brand pharmaceutical manufacturers have been identified as strategies designed primarily to hinder the more timely availability of cheaper generic copies of the brand-name medication:

1. *Launching next generation brand product lines*, such as extended release products, of the new drug whose patent is expiring;

2. *Initiating the sale of "authorized generics"* through lucrative contractual arrangements with cooperating generic drug manufacturers;

3. Contracting with and *compensating potential generic drug manufacturer competitors to delay the introduction of first-time generic products* on the market;

4. *Selling the brand-name company's own generic version* of its patent-expiring brand through a company-owned generic subsidiary;

5. *Expanding clinical indications to extend patent protection*; and

6. *Filing last minute or frivolous citizens petitions* to delay the FDA's approval of a new generic for marketing.

These controversial major anti-competitive, anti-consumer practices of the brand-name drug industry will be discussed below. Proponents of the individual practices will defend each as being in the interest of the paying consumer. Likewise, opponents are loud in their damaging assessments of the same as being anti-competitive and harmful to patient access to affordable or safe drug therapy or both. The effects of these concerns on consumer drug affordability and safety are also discussed.

Using "Authorized Generics" to Foil First-Time Generics

Contracting with generic pharmaceutical manufacturers to offer *"authorized generics,"* at the beginning of the legally protected 180-day period of exclusivity for the winner of a drug patent challenge, has become a routine practice in the industry. The legislation creating the period of exclusivity was enacted to provide incentives for the first generic drug manufacturer to successfully challenge the patent of the brand-name pharmaceutical company to be able to market its first-on-the-market generic without competition from other generic firms for a period of six months after expiration of the innovator drug's patent. The objective of the original legislation was to encourage generic drug manufacturers to bring new generics to the market more quickly in order to promote competition and bring about lower drug prices for consumers.

The practice of providing "authorized generics" is contrary to the original intent of this legislation and is one of the most controversial actions taken by the original drug patent holder to preserve revenues from the higher prices charged for their single-source, brand-name prescription drugs. This practice is inimical to consumers because it can adversely affect generic drug manufacturers' motivation to seek first-generic status after patent expiration of a single-source drug. In so doing, this practice tends to foil the intended 180-day period of exclusivity for the first generic manufacturer before other generic firms are allowed to compete with them.

Because of Big Pharma's invention of "authorized generics," consumer access to significantly lower cost generic pharmaceuticals can be negatively impacted and losses in potential savings can be substantial, in fact, in millions of dollars annually.

According to the GPhA, authorized generics are merely brand-name drugs "masquerading as generics" and are generally marketed only during the 180-day period of exclusivity awarded by the FDA to the first generics company to have successfully challenged a single-source manufacturer's questionable drug patent. At the end of the 180-day period of exclusivity, any pharmaceutical manufacturer with an FDA-approved product can market the generic drug, providing further competition to keep consumer prescription drug prices lower.

To counter these arguments, PhRMA's issued a statement on authorized generics, backed by a PhRMA-sponsored study conducted by IMS Health, to defend its position. The conclusions of the report, "Assessment of Authorized Generics," were that authorized generics actually foster competition and result in lower prescription costs. Accordingly, the average generic price discount to brand at pharmacy outlets is about 16 percentage points greater than would be with comparable products in the absence of an authorized generic drug product during the 180-day period of exclusivity. Beyond this period, they argue, the differences in prices comparing authorized generics to regular generic drugs depend on the number of generic competitors on the market.

Big Pharma promotes the view that authorized generics are a win-win situation for consumers by increasing competition and reducing drug prices and overall health care costs. While they assert they do not determine the final retail drug prices to the consumers, the manufacturer prices to the dispensing retail pharmacy are the major contributor to consumer prescription costs at the pharmacy. These manufacturer prices usually account for almost three-quarters of the final price of a brand-name prescription drug to the consumer, according to industry intelligence.

Because of its importance to consumers and the taxpaying public, the issue of major pharmaceutical manufacturers consistently offering "authorized generics" has become sufficiently hot that the Federal Trade Commission (FTC) was requested and has taken steps to investigate whether and to what extent such a practice is anti-competitive. Many believe such practices are anti-competitive and, therefore, anti-consumer, and that they seriously threaten drug price affordability. The senators requesting this investigation asked the FTC to look into:

The short-term competitive benefits of "authorized generics" during the FDA-approved 180-day period of market exclusivity.

1. The long-term impact of the practice of "authorized generics" on competition in the drug market.

2. The price of drugs.

3. The viability of the generic drug industry.

In announcing its intention to conduct this probe, the FTC said it would examine the following:

1. Actual wholesale prices (including rebates, discounts, etc.) for brand-name and generic drugs, both with and without competition from authorized generics.

2. Business reasons that support authorized generics entry during a first-filers' 180-day period of exclusivity.

3. Factors relevant to the decisions of generics firms about whether and under what circumstances to seek entry prior to patent expiration.

4. Licensing agreements with firms producing authorized generics.

Brand/Generic Settlements Delay First-Time Generics

Innovator pharmaceutical firms have also been increasingly contracting with and compensating generic firms to delay the launching of first-time generic copies of their brand-name product for years. During such time periods, the brand pharmaceutical firm continues to amass voluminous sales from their much higher priced single-source products. Unfortunately, during this time period consumers are robbed of any potential savings they would enjoy by having a cheaper, generic form available.

The Federal Trade Commission (FTC) is the federal watchdog agency charged with policing anti-competitive practices in business. Several of its pro-consumer decisions have been reversed on challenge in federal courts with some recent decisions deeming the settlements to be legal.

This practice has also become more commonplace in recent years. One of the most highly publicized cases of a major pharmaceutical company contracting to pay a generic pharmaceutical company, Apotex, to delay introduction of its generic product on the U.S. market is the BristolMyersSquibb/Sanofi-Aventis/Apotex furious legal episode involving the blockbuster clot buster drug, Plavix (clopidrogel bisulfate). Litigation surrounding this case involved multiple legal twists and turns which illustrated several industry brand defense mechanisms at play in the one case.

The FTC's Bureau of Competition recently issued its Summary of Pharmaceutical Company Settlement Agreements for FY 2006. This report is mandated by the Medicare Prescription Drug Improvement and Modernization Act (MMA) of 2003. MMA requires pharmaceutical companies to file certain agreements with the FTC and the U.S. Department of Justice.

This report clearly showed an escalation of pharmaceutical industry settlements after FTC's pro-consumer decisions against a Big Pharma company were overruled in court. Of 28 final settlements in fiscal year 2006, half of the generic companies both received compensation and agreed not to market its product for a period of time. Compare this with the activity of the previous two years. In 2005, only three of the eleven settlements in 2005 had both provisions while none of the 14 settlements in 2004 had both compensation and generic marketing restriction provisions.

The compensation provided for in the settlements to the generic pharmaceutical companies took different forms. These included:

1. Payments for co-promoting the brand-name product;

2. Payments for supplying, or being available to supply, the brand-name company with raw material or finished drug product;

3. An agreement by the brand-name producer not to compete with an authorized generic;

4. Payments for intellectual property to the brand; and

5. Payments as part of a co-development project between the brand and the generic.

In the world of the generic pharmaceutical industry, a first-filer is the first generic manufacturer to file an ANDA that claims the patent (or patents) protecting the brand drug are invalid or will not be infringed by the generic's product. As an incentive for first-filers to bring less costly generics to market as soon as possible, federal law provides for 180 days of market exclusivity. During this period, the Food and Drug Administration may not, with few exceptions, approve another generic filer's product until 180 days after the first-filer generics company goes to market. As a consequence of the court's earlier decision overruling FTC's pro-consumer actions, nine of the 11 settlements involving first-filers contained both a payment to the generic and a restriction on generic entry.

Selling Own Generic Version of Own Brand Product

Several of the major brand pharmaceutical firms have their own generic subsidiaries. This enables them to offer their own generic versions of their own single-source prescription brand drugs. By doing so, the drug patent holder can sell its own brand-name drug in a generic version simultaneously with the beginning of the FDA-approved first filer's official 180-day period of market exclusivity. This enables the innovator company to benefit from this practice in several ways. For instance, it can:

1. capitalize on its own generic producer's name recognition to continue to garner sales after patent expiration for the sole source drug;

2. take advantage of its lower costs by not having to file an abbreviated NDA as required of competitors because its already approved NDA covers the identical formulation used in its own generic product;

3. set the generic's price closer to the brand price, resulting in higher prices to the patient and to discourage generic substitution by would-be competitors;

4. "punish" a first-generic pharmaceutical manufacturer for cutting short the innovator drug company's high price single-source drug monopoly; and

5. benefit from the generic's sales as revenues from its branded product lines decline due to upcoming generic competition.

Expanding Clinical Indications to Extend Patents

Applying for approval of new and extended clinical indications by the FDA for an existing brand-name drug involves satisfying the federal agency that the drug is useful and can be safely used in patients for additional diseases or conditions or in ways not previously approved in the company's NDA. It may involve obtaining approval for "age group extensions" as for use of the drug in younger patients, which is advantageous from a clinical perspective, since most new drugs were studied in and approved for use by adults, but not for use in children. Of course, this practice can produce significant public benefits, on the one hand or, conversely, serve primarily as a mechanism for high profit-retention by the drug's manufacturer. FDA approvals for additional therapeutic uses or age group extensions, such as for pediatric uses, can both add important patient care alternatives as well as effectively extend patent protection for the brand by up to three additional years.

Filing Frivolous Citizen Petitions to Delay Generic Drugs

Citizens petitions is a means of allowing the public to register opposition to the approval of drugs for various reasons. Various public interest groups and corporations can use this process to delay the approval of first-generic copies of a single-source branded product after patent expiry, when all other efforts fail or in addition to the use of other delaying strategies discussed in this section. That these filings are simply being used to delay the appearance of first-generic drugs on the market is clear from the approval results attained over the past 5 to 10 years. For instance, out of 21 citizens petitions filed to stop new generics from coming to market, 20 were found to be frivolous or without merit. Months or years of delay for a new generic drug can mean millions, even billions, of dollars in sales for an innovator company's blockbuster brand-name drug and hundreds of thousands to millions of dollars in lost consumer savings.

In addition to the above practices, Big Pharma companies engage in a variety of other strategies to protect their brand sales and profits.

Other Controversial Drug Industry Practices and Issues

The major pharmaceutical manufacturing industry players engage in other controversial practices which affect patient safety and/or promote potentially excessive demand for their products without providing adequate information on associated risks. Chief among these practices are their failure to complete post-market safety surveillance studies on their new drugs as promised; their heavy advertising of their single-source, patent-protected prescription drugs directly to consumers, in addition to both appropriate and inappropriate promotion of these to physicians; and their heavy lobbying of government officials at both state and national levels, the costs for which are ultimately passed on to the consumer.

Neglecting Post-Market Drug Safety Studies

In order to gain approval by the Food and Drug Administration (FDA) of a pharmaceutical company's new drug application (NDA), the company must provide scientific evidence that the drug is effective for the disease or condition it is to be marketed for (drug effectiveness) and safe to use under the approved labeling (drug safety) and in the pharmaceutical dosage forms in which the drug has been tested clinically in humans. For initial approval by the FDA, new drugs are usually tested in only four to five thousand patients.

Patient dosing and adverse drug effects experienced by test subjects are determined and included in materials in the NDA submitted to the FDA. However, some side-effects may be too rare to be emerge in the relatively small groups of patients used in clinical studies, although such rare side-effects may be dangerous or even fatal when they do occur. In order to gain approval of their NDAs, companies agree to conduct post-market safety studies to amass further information and uncover such rare adverse effects, if they exist, and report this information to the FDA.

Once the drug is marketed, it will usually be used by hundreds of thousands—if not, millions—of patients during the early years of marketing. In such large populations, even the more rare side-effects are likely to emerge. These data, along with incidents voluntarily reported by doctors, pharmacists, patients and others, round out the information the FDA needs to ensure the long-term safety of the drug in medical practice.

The manufacturer can voluntarily initiate or at the FDA's behest be required to issue additional cautionary information and warnings regarding the use of an approved drug, in order to protect users from newly discovered risks. Therefore, failure of the manufacturer to conduct post-market follow-up safety studies exposes patients to needless risks of possible irreparable damage or even death, as has been amply illustrated by some of the more recent high profile drug product liability cases.

In some cases, severe warnings referred to as "black box warnings" may be required to be added to product labeling, both before and after the initial FDA approval of a new drug. This effectively limits the use of the drug to a narrower, more specific range of patients. Black box warnings highlight serious information about prescription drugs such as adverse drug reactions (side-effects, drug interactions, allergies, etc.), dosing information, at risk groups, and special requirements for taking, administering and monitoring use of these drugs. Black box warnings have been applied to hundreds of prescription drug products across a broad range of therapeutic categories.

In still other cases, emerging safety data may show a potential for such severe harm to users of the drug that either the FDA outright bans further marketing of it or the manufacturer voluntarily removes the drug from the market.

As recently as 2006, such serious lapses in reporting harmful patient safety data were front page headlines as a result of high visibility drug liability cases involving a major pharmaceutical manufacturer's recently withdrawn non-steroidal anti-inflammatory blockbuster drug Vioxx by pharmaceutical giant Merck.

As a result of such incidents, legislation was recently introduced in Congress to empower the FDA to compel pharmaceutical manufacturers to conduct their post-market safety followup studies to prevent the recurrence of such incidents. Finally, as this book was going to press, Congress approved the legislation to strengthen the power of the FDA to enforce the post-market drug safety requirement on pharmaceutical manufacturers.

Issued on the last day of Summer 2006, a scathing report from a government advisory group, the Institute of Medicine, comprehensively addressed this major safety issue. This report, addressed the following specific problems relating to the underlying causes and proposed solutions to this vital issue:[51]

1. Improved labeling requirements and limits on advertising for new medications.

2. Additional enforcement tools for the FDA.

3. Clarification of the authority of the FDA.

4. Clarification of the FDA's role in gathering and communicating additional information on the risks and benefits of marketed drugs.

5.   Mandatory registration of clinical trial data to improve public access to information on drug safety.

6.   Expanding the role of the drug safety staff of the FDA.

7.   Substantially boosting both funding and staffing of the FDA.

Promotion of "Off-label" Uses of Drugs by Industry

To gain approval for a new drug, the pharmaceutical manufacturer must prove the effectiveness and safety of the new drug product usually through randomized, double-blind, placebo-controlled clinical trials in patients who have the disease condition for which the drug will be marketed. After approval of the NDA, the drug manufacturer is legally authorized to promote its product to prescribers only for the indications and circumstances of use documented in the approved NDA. The NDA holder is prohibited by law from promoting its product for other medical uses without first gaining approval of the FDA by submitting supplemental data proving its efficacy and safety for use in the newer indications.

Drug labeling includes the approved manufacturer's printed package insert and all other packaging for the particular drug product. An "off-label" use in clinical practice is the use of the drug in a manner or for an indication the FDA has not approved. Consequently, such use is not included in the FDA-approved labeling for the product. Manufacturers can legally claim—and, therefore, promote—only those uses in their approved NDAs.

However, the off-label use of drugs is commonplace, contrary to widespread misconceptions, as shown by data from a recent government-funded survey of office-based physicians. According to data from the study, 21 percent of the 725 million prescriptions written by the responding doctors were for off-label, or unapproved, uses. Of these, nearly three fourths were reported to lack adequate scientific backing for their off-label use, but rather were based on observation or case reports, rather than on carefully documented clinical studies. As asserted by one of the study authors, while "some off-label uses are logical extensions of the FDA-approved drug labeling ... for others, clear evidence pointed toward experimentation."

The law prohibiting pharmaceutical companies from promoting the use of their drugs to practitioners for off-label uses places no such prohibition on the practicing physician's use of the drugs for unapproved indications. In other words, practicing physicians are free to use approved drugs for off-label uses routinely in their practices as they see fit so long as it is for the benefit of their patients.

There are many reasons physicians use approved drugs for off-label or unapproved uses. First and foremost, the freedom to employ off-label uses of drugs is deemed necessary for optimal treatment of many of their patients.

Desperation to find a treatment for patients after all approved drugs for the condition have been tried, as is standard for treatment of cancer and AIDS patients, is a potent reason for doctors to resort to off-label uses of drugs. Similarly, to handle intractable medical problems is another. For instance, the successful use of the Alzheimer's drug Aricept® to treat an antipsychotic drug-induced movement disorder disability in a 16-year old is an amazing example. Necessity in the case of children is often the compelling force for doctors to resort to off-label use of drugs as most drugs on the market are approved for use in adults but not in children, for obvious reasons.

The off-label use of drugs is also deemed necessary for the discovery of new uses for old drugs. In fact, some 57 percent of drug innovations are reported to have been discovered by physicians through off-label uses in their medical practices as opposed to new use discoveries by the pharmaceutical industry or academic researchers.

One of the more recently reported questionable, though apparently effective, off-label uses of prescription drugs is the use of Inderal (propranolol) to rid musicians of stage fright during their performances.

On the other hand, the off-label use of drugs can be risky or harmful when used inappropriately. One such example is the widespread practice of prescribing antibiotics for the common cold. Many patients demand antibiotics from their doctors for their colds. Antibiotics are ineffective against viral colds, subject patients to risks of allergic reactions to the antibiotics prescribed and contributes to an already serious global problem of antibiotic resistance. Contrary to scientific evidence and warnings from the CDC, many consumers consider the use of antibiotics to treat their colds as good medicine.

The off-label use of estrogen for decades for a wide range of female medical problems should provide some insight into how off-label uses of drugs can be harmful. Finally, in 2002, estrogen was reported to actually increase the risk of several serious diseases, including heart disease, stroke and breast cancer.

Gabapentin (Neurontin, Warner-Lambert, a subsidiary of Pfizer) is perhaps the best example of recent cases where the manufacturer flagrantly promoted the use of a prescription drug for non-approved, off-label uses over an extended period of time. When sued for these infractions, the company agreed to plead guilty and pay criminal and civil penalties totaling some $430 million.[52]

Neurontin, initially approved by the FDA in 1993, was launched on the market in early 1994 solely for use along with other drugs to control epileptic seizures. According to the government, the company aggressively marketed Neurontin for solo treatment of persons with epilepsy as well as for a long list of disorders not approved in the original NDA.

The list of unapproved conditions the company promoted to physicians for Neurontin treatment included bipolar mental disorder, various pain disorders (diabetic neuropathy, etc.), Lou Gehrig's disease, attention deficit disorder, migraine, drug and alcohol withdrawal seizures, and restless leg syndrome.

A 2003 analysis of off-label prescription drug use showed that fully 90 percent—or $1.8 billion—of Neurontin's sales were for uses classified as off-label or unapproved. An illustration of the scope of off-label use of Neurontin compared to a variety of other drugs employed in off-label treatments is presented below:[53]

1. *Anti-seizure drugs* Neurontin, Topamax and Depakote:

    a.   Neurontin, 90 percent off-label uses on sales of $1.8 billion

    b.   Topamax, 79 percent off-label uses on sales of $643 million

    c.   Depakote, 25 percent off-label uses on sales of $162 million

2. *Antibiotics* Biaxin, Avelox, Levaquin, Zithromax and Cipro:

   a. Biaxin XL, 58 percent off-label uses on sales of $173 million

   b. Avelox, 56 percent off-label uses on sales of $128 million

   c. Levaquin, 42 percent off-label uses on sales of $446 million

   d. Zithromax Z-Pak, 37 percent off-label uses on sales of $409 million

   e. Cipro, 30 percent off-label uses on sales of $291 million

3. *Antidepressants* trazodone, Remeron and Wellbutrin products:

   a. Trazodone, 56 percent off-label uses on sales of $120 million

   b. Remeron, 46 percent off-label uses on sales of $83 million

   c. Wellbutrin SR, 27 percent off-label uses on sales of $464 million

4. *Antipsychotics* Seroquel, Risperdal and Zyprexa products:

   a. Seroquel, 78 percent off-label uses on sales of $778 million

   b. Risperdal, 65 percent off-label uses on sales of $979 million

   c. Zyprexa, 42 percent off-label uses on sales of $913 million

The "fen-phen" combination was prescribed off-label for weight loss in order to take advantage of opposing side-effect profiles for the two drugs fenfluramine and phentermine. Phentermine was approved for marketing in 1959 and fenfluramine was approved in 1973. Both drugs had been approved for short-term treatment of obesity for a few weeks only. However, some physicians were prescribing the two drugs in combination for extended periods of time for weight loss programs.

Their combined off-label use was responsible for skyrocketing prescription sales of fenfluramine from 100,000 prescriptions per year to 5.1 million annually within a few years.[54] In 1997, the Mayo Clinic reported that 24 "fen-phen" patients had developed heart valve disease and pulmonary hypertension.[55] Evidence suggested that the two drugs making up the fen-phen combination caused heart valve disease, primary pulmonary hypertension and neurotoxic brain injury. At FDA's urging, the manufacturers of fenfluramine (Pondimin) and dexfenfluramine (Redux), both subsidiaries of American Home Products, withdrew both drugs from the market in September 1997.

Other classes of medications commonly prescribed for off-label uses are heart, asthma, allergy and seizure drugs and can come from just any therapeutic class of drugs. Unless effectiveness and safety are in question, no one can or should interfere with the physician's professional prerogative to use of drugs for off-label uses when deemed necessary to protect their patients' health and welfare. Consumers, however, have a right to raise issues and obtain explanations from their prescribers for uses of their prescription drugs, especially when the uses are different from those in the FDA-approved product labeling.

Direct-to-Consumer (DTC) Advertising

For years, the major pharmaceutical manufacturers have used the media to bombard consumers with heavy advertising to promote the use of their brand-name prescription drug products. The industry defends these practices as being patient or consumer education about what treatments are available. In practice, however, these massive DTC campaigns encourage consumers to discuss the advertised drugs with their doctors on their first subsequent visit and many consumers will pressure their doctors to prescribe the DTC-promoted product whether or not the drug is best for them.

According to the World Health Organization, the United States and Australia are the only countries where DTC is allowed. For more than a decade, U.S. pharmaceutical manufacturers have been allowed to advertise prescription drugs directly to consumers as well as to doctors and other health care professionals. Because of the potential for abuse by presentation of incomplete and unbalanced information, especially safety information, and the potential for consumers to demand prescriptions for medications which may be inappropriate for them, this industry practice continues to be highly controversial. In view of several studies and recently adopted industry guidelines for DTC advertising, drug consumers can empower themselves to better evaluate the appropriateness of their reliance on DTC ads for impartial drug information.

While brand-name drug manufacturers insist that DTC advertising provides a valuable educational service for consumers, increases in demand for the products advertised directly to consumers correlate well with company DTC advertising expenditures. Also, some prescribers feel that this practice serves a useful purpose but may also come up short in communicating balanced information about drug safety risks in these ads.

Heavy Industry Lobbying Protects Corporate Interests

Heavy investments in lobbying Congress and states by national organizations representing the major brand prescription drug manufacturers are rewarded with legislation favorable to their interests, such as strong patent protection, direct-to-consumer advertising, or inaction on legislation unfavorable to their interests, such as drug importation and price controls. For example, the Pharmaceutical Research and Manufacturers of America (PhRMA), which represents all of the major global pharmaceutical manufacturers and their subsidiaries, is reported by The Center for Public Integrity to have spent some $74 million in lobbying expenditures since 1998.

The Biotechnology Industry Organization (BIO), a similar group representing manufacturers of biopharmaceutical products (including some of the largest members of PhRMA), is reported to have invested $23.6 million in lobbying expense since 1998.

Each of the top 20 brand-name pharmaceutical manufacturers individually incur huge lobbying expenses in addition to those spent by the two industry groups above. As an example, Pfizer, Inc., the largest pharmaceutical manufacturer is reported to have spent at least $43.5 million on lobbying from 1998 through 2005; with GlaxoSmithKline (GSK) following with about $32.4 million since 1998.

While overall philosophies may differ between the two major political parties, the tactics of the pharmaceutical industry remain virtually the same, irrespective of the ruling party in Congress. The recent changeover in party control of Congress most perfectly illustrates the power of the pharmaceutical industry (PhRMA) lobby over the legislative branch of government in the United States. Upon assuming party control of Congress in January 2007, the Democratic leadership had announced their intention of proposing a new government-run, rather than private, prescription drug program for seniors, as a means of reducing prescription drug costs associated with the new Medicare drug benefit first initiated under a Republican-controlled Congress, beginning early 2006.

Because of continued strong support by the Bush Administration coupled with the tremendous power still wielded by the formidable drug lobby, even in a Democratic-controlled Congress, the new House leadership settled for a simpler, less ambitious plan in order to complete their initial "100 hours" legislative agenda. Rather than pursue the original, more complex government-run option, the party chose the option of requiring the government to negotiate directly with pharmaceutical manufacturers for lower prescription drug prices within the existing Medicare drug plan.

According to a 2007 article in the Washington Post, the industry trade group PhRMA spent "more than $1 million on full-page ads touting the success of the existing Medicare drug system" in January alone.[56] While about three-quarters of some $90 million donated to federal candidates and party committees during the period 1998 to 2005 went to Republicans, the pharmaceutical industry lobby now appears to be rapidly increasing contributions to Democratic leaders and increasingly employing Democratic lobbyists to facilitate access to the new committee chairpersons.

It is, therefore, important for the electorate to become more fully aware of these powerful influences on both the executive and legislative branches of our government and how these influences translate into continued protection of the interests of the pharmaceutical industry over those of our tax-paying consumers.

Regulatory Issues Affecting Drug Safety and Costs

A number of practices and issues limiting the FDA's ability to fulfil its mandate to protect the public and to expedite the availability of affordable and safe drug therapy to consumers are cause for great concern of consumers and health care providers, as well as national political leaders. Among the most prominent of these are the following:

1. The absence of a procedure for speeding the availability of less costly generic copies of biologicals or biopharmaceuticals, by far the most expensive category of prescription drugs;

2. A huge backlog of applications for approval of generic drugs, the least expensive category of prescription drugs;

3. The increasing threat of counterfeit and stolen drugs entering a surprisingly safe but vulnerable drug supply chain in the United States; and

4. User fees, as opposed to government funding, used as principal mechanism for speeding up review and approval of new drug applications from the pharmaceutical industry.

Huge Backlog of Generics Waiting for Approval

Even as the floodgates from a virtual bumper crop of prescription drug patent expirations open to allow a rapid expansion of new generic drugs on the market within a relatively short period of time, the FDA finds itself with a backlog of a historically high number of generic drug applications due to inadequate funding. As of the beginning of 2006, the agency's backlog was at around 800 unprocessed applications. Reducing this horrendous backlog will greatly expand the number of cheaper but equally effective, generic substitutes available to the consumer.

On the one hand, the increased availability of generic drug products will allow consumers to realize substantial savings in filling their prescriptions. On the other hand, the implementation of generic drug user fees will naturally increase the prices of these generics to consumers as they come to market. This matter is one of great importance in light of annually increasing prescription drug costs and overall health care costs.

Lack of Biological Generics Approval Process

Because of the potential for profit from the more expensive class of pharmaceuticals—the biopharmaceuticals or biologics—as well as the impact that increased availability of generic pharmaceuticals has had on the brand-name drug industry recently, it comes as no surprise to find the major pharmaceutical industry's gradually changing the focus of their discovery, and research and development efforts to this newer class of pharmaceutical compounds. Based on the limited activity involving the processing of biological generics or "bio-similars", the costs for even generic forms in this class could still be expected to be prohibitively high because of the complexity of the therapeutic molecules making up these drugs. Couple these reasons with the glaring absence of a standard process for facilitating the FDA approval of generic forms of biological pharmaceuticals, and the potential for a new round of escalating health care costs, pushed by these newer bio drugs, can be envisioned before relief from effective competition from biogenerics will be able to moderate overall prescription drug prices beyond that projected from ordinary prescription drug patent expirations over the next several years.

Our political leaders have a responsibility for ensuring that appropriate legislation and funding provide for adequate manpower and appropriate procedures and processes within our national regulatory agency—the FDA. The agency needs to have the power, the mandate and adequate funding to protect the public health, in this instance, by ensuring the future access to affordable biogeneric medications, which are much more costly to develop and produce.

Counterfeit Drugs and Consumer Safety

The extent of fraudulent or counterfeit drugs in North America—the United States and Canada—has been estimated to be in the range of 1 to 2 percent of drugs. With current growth estimates of prescription drug counterfeiting, worldwide drug counterfeiting is projected to rise from the recent $35 billion business estimate to an estimated $75 billion business by 2010. While the percentage of counterfeit drugs in North America may seem low to some, counterfeit drug sales are increasing at nearly twice the rate for sales of legitimate prescription drugs. According to the FDA, between 2004 and 2005 alone, customs officials confiscated nearly twice as many fraudulent prescription drugs.

Drug counterfeiting threatens consumer lives by interfering with the safety of the nation's drug supply. Drug counterfeiters pose serious threats to the safety of drug products flowing through pharmaceutical supply chain from manufacturer to the ultimate consumer, with special vulnerabilities through transactions between secondary and primary drug wholesalers, as well as in consumer drug purchases from non-regulated internet pharmacies and foreign mail order sources. Even in well-intentioned purchases of less expensive prescription drugs from Canadian pharmacies, the door to increased entry of fake drugs is widened even further.

While counterfeit drugs may come from almost anyplace, two of the most common origins are China and India, according to official sources. Because of yet unplugged holes in our nation's drug supply chain, consumers can have prescriptions filled with counterfeit drugs even at reputable pharmacies and, most especially, through illegitimate but legitimate-looking Internet pharmacies.

The most serious influences in increasing criminal drug counterfeiting are an expansion of under-regulated drug wholesalers and re-packagers, the increasing number of pharmacies on the Internet, and the continually increasing importation of prescription drugs from other countries.

Counterfeit drug products intercepted by the government have been found to be composed of crude mixtures of glue, chalk and sugar, as well as virtually any kind of fabrication in between these and products difficult to distinguish from the real thing. The appearance of some counterfeit drugs makes their fraudulence obvious or at least suspicious. For instance, some telltale signs of fake medicines include differences in color, taste, shape of pill (tablet, capsule, etc.), lack of appropriate identification on pill, packaging, labeling, and unanticipated side-effects. The pharmacist at the local pharmacy is the most accessible avenue for aiding consumers in determining if their medication is fake or fraudulent. If strange side-effects are experiences after taking a questionable drug product, consumers should contact their doctors, the nearest poison control center or their local pharmacist.

In order to plug the holes in our nation's prescription drug supply chain, a repeatedly delayed national pedigree program was scheduled to start by January 1, 2007, under the aegis of the FDA, the federal drug regulatory agency, and the state boards of pharmacy, which regulate and license pharmacists, pharmacies and wholesale drug distributors in each state. This program is designed to provide an unbroken chain of custody of prescription drug products from the original pharmaceutical manufacturer's warehouse through one or more wholesale drug distributors and re-packagers of drugs obtained in large quantities in bulk, where applicable, all the way to the local community or hospital or mail service pharmacy where they are dispensed to waiting patients.

Drug counterfeiters focus on selling fake copies of high value drugs. Consequently they are prone to produce fake copies of drugs most beneficial to their profit motive: top selling drugs in high demand, the most expensive drugs, and drugs for some of the most serious diseases or health conditions. Also, since their motives are purely profit-driven, there is absolutely no concern for the lives of consumers who obtain and take their fake products.

Because of the high profit to be made from selling counterfeit drugs, the criminal element provides virtually exact chemical replicas of certain complex pharmaceutical drugs, including the most expensive biopharmaceuticals such as epoetin, a biosynthetic anemia drug; Lipitor (atorvastatin), cholesterol-lowering drug; and Viagra (sildenafil), a drug for erectile dysfunction. A more complete listing of such products intercepted over the past few years is found below. Lipitor has been the top selling drug in the world for several years and would be a likely target of such counterfeiters. Epoetin is an expensive biopharmaceutical drug product, also fitting the counterfeiters' selection criteria. These along with other counterfeit products make up the following recent national list of susceptible prescription drugs:

1. Combivir (lamivudine/zidovine) for the treatment of HIV/AIDS.

2. Diflucan (fluconazole), an antifungal agent.

3. Norvasc (amlodipine), an anti-hypertensive agent.

4. Epivir (lamivudine) for HIV/AIDS.

5. Epogen (epoetin alfa), a biosynthetic for the treatment of severe cases of anemia.

6. Evista generic (raloxifene) for osteoporosis in postmenopausal women.

7. Lamisil (terbinafine) for treatment of fungal infection.

8. Lipitor (atorvastatin) for lowering cholesterol.

9. Serostim (somatropin), a biosynthetic human growth hormone.

10. Procrit (epoetin alpha), a biosynthetic for the treatment of severe cases of anemia.

11. Trizivir (abacavir/lamivudine/zidovine) a three-drug combo for the treatment of HIV/AIDS.

12. Viagra (sildenafil), for erectile dysfunction.

13. Zerit (stavudine) for HIV/AIDS treatment.

Needless to say, such a list as this one remains in flux and is subject to change periodically as new counterfeit prescription drug products are found and confiscated by federal authorities. Consumers can obtain information on the most current listing of counterfeit drugs by calling or checking the Web sites of the FDA at *www.fda.gov* or their respective state health departments, especially the state boards of pharmacy. Consumers should also report to their pharmacy any products they receive and suspect to be counterfeit. They may also contact state health officials.

Chapter 8: Generic Drugs—The Pro-Consumer Alternative

Consumers are properly concerned about prescription drug prices and their out-of-pocket costs, particularly since most are either unaware of or bewildered by the complex forces and factors influencing or directly affecting their out-of-pocket expense.

While the annually increasing share of overall health care costs attributable to prescription drug costs has dramatically decreased over the past few years, prescription drug costs continue to rise at rates faster than the rate of inflation. The major burden in these costs is the significantly higher prices for brand-name prescription drugs, which overwhelmingly account for the major share of national drug expenditures. Generic drugs are, however, a pro-consumer alternative to expensive brand-name prescription drug products, once these products lose their patent protection. The legal authorization and expansion of generic drug competition in medicine has proved to be the single most important factor in reducing and moderating overall consumer drug costs.

An explosive pro-consumer generic drug revolution is now underway in the United States due to a large number of major brand-name prescriptions going off patent between 2005 and 2010. However, as discussed in a preceding chapter, powerful anti-consumer, anti-competitive forces are hard at work in attempts to impede this consumer-positive trend to the extent possible. The discussion of the issue is continued in this chapter as we explore the extremely slow movement toward a new level of generics development: biogenerics, biosimilars, or generic biopharmaceuticals. The implementation of a safe but economical process for providing generic substitutes for biopharmaceuticals would further reduce consumer and government drug spending greatly as patents expire for the oldest drugs in this newest, most advanced, and most expensive class of prescription drugs.

The proportion of prescriptions filled with generic drugs has been steadily growing, and the acceleration continues principally due to government efforts to pare its overall prescription drug costs and overall health care costs for programs it runs for its various constituencies. Such constituencies include enrollees and beneficiaries in Medicaid and Medicare, programs for veterans and their families, the military and their families, and other programs funded in whole or in part by governmental agencies at both the federal and state levels. The result is that all consumers benefit from lower cost, equally safe and effective generic copies of previously patent protected, single source brand-name products.

Fully 70.4 percent or 1.254 billion of the 1.781 billion prescriptions filled with regular, unbranded generic drug manufacturers were produced by the top 10 generic prescription drug manufacturers during 2005. Generic drug prescriptions represented 56 percent of the total prescriptions dispensed in the United States during 2005, but only about 13 percent of all dollars spent on prescription medications, according to IMS Health data. Further, generic drug use was estimated to grow by some 13 percent in 2006.

Recent data released by the Centers for Medicare and Medicaid Services (CMS) showed 56.9 percent of drugs dispensed to enrollees in Medicare prescription drug programs and Medicare Advantage plans through third quarter 2006.[57] Among private third party payers, generic drug dispensing increased by 9 percent from 2005, when it was 48.4 percent, to 2006 at 52.6 percent, according to the National Association of Chain Drug Stores (NACDS).

What Makes a Drug Generic?

Generic drugs are simply non-patented FDA-approved copies of an innovator company's brand-name drug, produced and marketed after innovator's products lose their patent protection. While overwhelmingly marketed under their generic names only, generic drugs may be marketed under their own unique brand names. Prior to the expiration of a drug manufacturer's patent, only that producer can market a new drug under either its brand name or its generic name except by license or contract from the patent holder.

Drugs bearing only the generic name are referred to as *unbranded generics*. By far, *most generics are sold under their generic names only*. Branded generics are merely generic drugs for which the manufacturer chooses to use a brand name to identify its product for name recognition.

The generic name is the official chemical name for the drug entity in the "pill" or pharmaceutical dosage form and is common to all brand-name drugs and their generic substitutes. It is the name by which all products containing the active constituent can be compared irrespective of whether a product is marketed or dispensed as a brand-name or generic drug product. All drugs can be referenced by their generic name whether protected by patent or not. The vast majority of generic drug products available are unbranded generics. Many companies which produce branded generics produce unbranded generics as well, depending on their chosen product mix for market positioning.

While all drugs have generic names—their common chemical names—and these must always appear on their product labels, generics can be marketed only after the pioneer drug manufacturer's drug patent expires. This is roughly 20 years from the date the pioneer drug was patented by the drug manufacturer, not the date the drug was initially approved for marketing in the United States. Federal law permits compensation for some of the patent protection time lost by a drug manufacturer during the research and development process after the patent is granted.

Presently, the U.S. consumer is witnessing a rash of patent expirations for an unusual number of blockbuster and regular branded prescription drugs. The rash of patent expirations on major brand-name drugs over the past few and upcoming years, along with an aggressive and highly competitive generic pharmaceutical industry, has strengthened generic prescription drugs as a major factor in reducing overall health care costs through lower prescription drug costs. This is a result of its slowing the annual rate of growth in consumer prescription drug costs dramatically, and guarantees that this trend will expand in the near future.

The impact of new generics on market share of brand-name drugs is often dramatic and swift, especially through the larger pharmacy chains and mail order pharmacies. With integral generic interchange programs, their large volume purchases and the power of widespread state-mandated generic substitution laws, retail pharmacies can achieve generic substitution rates of some 90 percent within the first month or so after a new generic is first introduced on the market, leaving only about a 10 percent or smaller market share to the higher priced, original single-source brand prescription drug.

Generic Drugs Are Consumer-Friendly

Answers to the following questions help one understand why generic drugs are the drug consumer's best friend:

1. Are generic drugs as safe as brand-name drugs?
2. Are they as strong as brand-name drugs?
3. Do they work as well as brand-name drugs?
4. Why do generic drugs cost less than brands?
5. How are generic drugs approved and regulated?

Generic drug products are required to be the same as their brand-name counterparts—same chemical, same potency, same purity, same dosage form—except for such physical characteristics as color, shape and size of the dosage form. Any change in the chemical identity or quantity of the active ingredient in a dosage unit, or dosage form makes the product a new drug—not a generic.

Generic Drugs Are as Safe as Brand-Name Drugs

Generic drugs are as safe as the brand-name products they substitute for because they are subject to the same standards of safety, purity and effectiveness as the pioneer brand. The generic product must contain the same active drug—not a modified form of the drug—in the same dosage strength, dosage form (capsule, tablet, syrup, etc.) and purity, and treat the same conditions in the same doses as the original brand-name product the generic is to replace. Inert or inactive ingredients in the product can vary according to the manufacturer so long as the active ingredient works the same. Further, many generic drugs are manufactured or sponsored by the same producers or by subsidiaries of the producers of the brand-name product itself.

Generic Drugs Work as Well as the Brands

To gain the approval of the FDA, manufacturer sponsors of a generic drug must prove their product to be bioequivalent to the innovator product it is to substitute for. Basically, this means that once the product is taken or applied, the active drug must release in the body in a manner and extent comparable to the drug's release from the original brand-name product it replaces. Otherwise, the request for approval of the "generic" product will be rejected. Again, the major brand-name pharmaceutical manufacturers also produce generics of their own innovator products to compete with generic versions from their competitors.

Most states require that generic drugs be generically equivalent to the brand-name counterpart. This means that they contain the same active drug and same strength in the same dosage form. Further, the generic must be the same chemically and produce the same medical effect when administered, applied or taken orally. In other words, it must also be therapeutically equivalent. Consequently, generic drugs provide the same risks and benefits as do their pioneer drug counterparts.

Generic Drugs Cost Less Than the Brands

Generic drugs can be marketed and sold for prices 30 to 80 percent less than their brand-name counterparts because their manufacturers are not required to make the heavy investments in research and development, promotion and marketing as the innovator prescription drug's manufacturer or sponsor. This is because the safety and effectiveness of the active drug have been adequately proven during the extended time of marketing enjoyed by the innovator company during its monopoly while enjoying patent protection from competition. Such periods can be up to 20 years for the average new drug.

The time-consuming and costly discovery and screening of thousands of drug candidates, nor the especially expensive clinical research studies, are necessary since the innovator company's research data are on file with the FDA. Since Big Pharma reports that the average time for a new drug to move from discovery through FDA approval is from 12 to 15 years and that it costs approximately $800 million to bring a new drug to market, the difference between investments of generic and brand-name drug manufacturers is substantial.

When Big Pharma companies produce their own generic versions of their branded drugs to compete with generic drug manufacturers, they use their already approved NDA, without any additional expense, to market them. Sometimes when a first-time generic is approved by the FDA for marketing, the single-source brand-name manufacturer may either contract with another generic drug manufacturer to market an "authorized generic" version of their branded product to compete with the legitimate first-generic product. Because "authorized generics" allow the use of the original brand-name drug manufacturer's already approved NDA, the price for the authorized generic will be lower than the price for the branded product. However, it will usually be higher than the price for a regular first-time generic because the original manufacturer shares in the revenues from sales of the "authorized generic" drug.

As a reward for their success in bringing a lower-priced generic to consumers, Federal law grants an initial 180-day period of exclusivity for the first generic pharmaceutical manufacturer who successfully challenges a drug patent and becomes the first ANDA filer to obtain FDA approval. Generic drug prices, though lower than the costs of the branded alternatives, are usually higher during this initial 180-day period of exclusivity for first-filers of generic drug applications. These first-generic producers will usually have the added legal expense of challenging the continued validity of the brand-name drug patent as well as other brand defenses engaged in by the innovator drug manufacturer in attempts to maintain its branded drug monopoly as long as possible. After this period of exclusivity, however, generic drug copies can be produced and sold by multiple manufacturers after obtaining FDA approval. This increases competition and increased competition further reduces generic drug prices to the consumer.

How Generic Drugs Are Regulated and Approved

Generic drugs are approved and regulated the same as new drugs are, with some modifications, by the same federal agency, the FDA. While new drugs require filing and approval of new drug applications (NDAs) prior to initial marketing, generic drugs require the filing and approval of abbreviated new drug applications (ANDAs). ANDAs are not required to contain the detailed clinical research data the full NDAs must include. However, all other information on procedures and processes must meet the same current good manufacturing practices as new drugs must.

For more detailed information on the FDA drug review and approval process, refer to Chapter 3 section on research and development.

The concept of therapeutic equivalence is of utmost importance in understanding generic drug substitution. For this reason, as a prerequisite to its approval of a generic drug product, the FDA requires evidence "that the generic drug is interchangeable with the brand-name drug under all approved indications and conditions of use."[58] In order to provide such assurances to prescribers, pharmacists and consumers, the FDA requires rigorous tests and procedures so that it is unnecessary for health care providers to perform additional testing when either of the following occurs:

- a brand-name drug product is switched to an approved generic drug product;
- an approved generic drug product is switched to the brand-name product; or
- one approved generic drug product is switched to another generic drug product when both are deemed equivalent to the brand-name drug product.

Measurements used variously, depending on the nature of the active drug ingredient when introduced into the body, to determine the interchangeability of generic drugs with brand-name drugs are determinations of *bioavailability* and *bioequivalence*.

Bioavailability is defined in federal regulations (§320.1) as:

> *"the rate and extent to which the active ingredient is absorbed from a drug product and becomes available at the site of action. For drug products that are not intended to be absorbed into the bloodstream, bioavailability may be assessed by measurements intended to reflect the rate and extent to which the active ingredient or active moiety becomes available at the site of action."*[59]

Bioequivalence is a concept involving comparison of one drug product with another and is defined in FDA regulations as:

> *"the absence of a significant difference in the rate and extent to which the active ingredient or active moiety in pharmaceutical equivalents or pharmaceutical alternatives becomes available at the site of drug action when administered at the same molar dose under similar conditions in an appropriately designed study."*[60]

In short, an applicant for FDA approval of a generic drug product must demonstrate that the product is *bioequivalent* to the pioneer brand-name product—that is, *the generic drug product must perform the same as the pioneer drug product*.

The Most Frequently Prescribed Generic Drugs

The twenty most frequently prescribed generic drugs in the United States during 2006 are listed below, along with their principal uses and reported sales:[61]

1. Hydrocodone/acetaminophen (APAP), for pain, $1,622,889,000.

2. Simvastatin (first-time generic), for cholesterol disorders, $1,390,479,000.

3. Zithromycin, antibiotic for infections, $1,216,982,000.

4. Oxycodone, for pain, $1,179,859,000.

5. Gabapentin, anticonvulsant and for neuropathic pain, $1,004,639,000.

6. Amoxicillin/potassium clavulanate, antibiotic for infections, $926,015,000.

7. Fexofenadine (first-time generic), antihistamine for allergies, $905,379,000.

8. Fentanyl transdermal patch, for pain, $840,242,000.

9. Lisinopril, for high blood pressure, $727,296,000.

10. Sertraline, antidepressant, $699,974,000.

11. Metformin, for diabetes, $670,167,000.

12. Omeprazole, stomach acid reducer, $642,235,000.

13. Clopidogrel, to prevent blood clots, $634,647,000.

14. Paroxetine, antidepressant, $621,809,000.

15. Fluticasone nasal, for treatment of asthma and allergic rhinitis, $616,191,000.

16. Amoxicillin, antibiotic, $492,911,000.

17. Albuterol aerosol, for asthma, $485,366,000.

18. Oxycodone/acetaminophen, for pain, $483,604,000.

19. Levothyroxine, for hypothyroidism and goiter, $471,735,000.

20. Lovastatin, for cholesterol disorders, $457,336,000.

Note that two of these most frequently prescribed generics were first-time generics—simvastatin, the generic for Zocor, and fexofenadine, the generic for Allegra.

The Top Generic Drug Manufacturers

Based on the number of prescriptions dispensed, four out of the top five U.S. pharmaceutical manufacturers were generic drug companies in 2005. These were Novartis (Sandoz), Teva, Mylan and Watson.

Based on the number of prescriptions filled with their generic pharmaceutical products, the companies below were the top ten generic drug manufacturers in the United States. The country of a company's headquarters and parent brand-name company, if any, are shown in parenthesis, along with the *total number of prescriptions filled* with products manufactured by each:[62]

- Teva (Israel) .... 234 million prescriptions
- Mylan (U.S.)... 213 million prescriptions
- Sandoz (Switzerland, Novartis) .... 191 million prescriptions
- Watson (U.S.) .... 158 million prescriptions
- Ivax (U.S.) .... 103 million prescriptions
- Mallinckrodt (U.S.) .... 84 million prescriptions
- Qualitest (U.S.) .... 74 million prescriptions
- Actavis (Iceland) .... 73 million prescriptions
- Par (U.S.)... 70 million prescriptions
- Barr Labs (U.S.).................................... 54 million prescriptions.

The top ten manufacturers with the *highest dollar sales for unbranded generics* in the United States during 2005 were:[63]

- Teva (Israel)*..$2,992 million
- Sandoz (Switzerland, Novartis) ....$2,643 million
- Mylan (U.S.)...$2,003 million
- Watson (U.S.) ....$1,425 million
- Ivax (U.S.) ....$1,186 million
- Par (U.S.)...$932 million
- Greenstone (U.S., Pfizer) ....$722 million
- Actavis (Iceland) ....$677 million
- Boehringer Ingelheim (U.S.)..................$674 million
- Baxter Healthcare (U.S.) ....$639 million.

Aggregate total dollar sales of generics for the top 10 manufacturers was $13,894 million out of a total of $22,317 million for the entire U.S. generic drug industry. In the preceding listing of top 10 marketers of generics in the United States, the country in parenthesis indicates the headquarters of the parent company; and the company in parenthesis, if any, indicates the parent company of the generics subsidiary listed. Teva and Ivax merged under Teva's name in early 2006.

First-Time Generic Drugs Approved in 2005 and 2006

Information about first-time generic drugs introduced on the market during 2005 and 2006 is detailed in Tables VII and VIII. Prior year sales for the 15 brands, for which first-time generics were introduced in 2005, totaled some $9.434 billion and included such blockbusters as Allegra for seasonal allergies ($1.296 billion), Duragesic Patch for chronic pain ($1.395 billion), Oxycontin for pain relief ($1.888 billion), and Zithromax, an antibiotic ($1.409 billion).

Table VII. FIRST-TIME GENERIC PRESCRIPTION DRUGS INTRODUCED IN 2005 WITH PRIOR YEAR BRAND SALES			
Brand Name and Dosage Form	**Generic Name of Drug**	**Medical Use**	**Prior Year Brand Sales**
Allegra tabs	fexofenadine	Seasonal allergies	$ 1,296,000,000
Amaryl tabs	glimepiride	Diabetes	$ 276,000,000
Arava tabs	leflunomide	Rheumatoid arthritis	$ 175,000,000
Biaxin tabs	clarithromycin	Bacterial infections	$ 198,000,000
Cefzil tabs	cefprozil	Antibiotic	$ 255,000,000
Copegus tabs	ribavirin	Hepatitis C	$ 200,000,000
DDAVP tabs	desmopressin	Diabetes insipidus; bed wetting	$ 172,000,000
Duragesic patch	fentanyl transdermal	Chronic pain	$ 1,395,000,000
Lamictal chew tabs	lamotrigine	Epilepsy and seizure disorders	$ 847,000,000
Oxycontin tabs	oxycodone	Chronic pain relief	$ 1,888,000,000
Sporanox caps	itraconazole	Fungal infections	$ 132,000,000
TriCor tabs	fenofibrate	Hyperlipidemia	$ 712,000,000
Ultracet tabs	tramadol/ acetaminophen	Acute pain relief	$ 331,000,000
Zithromax tabs	azithromycin	Antibiotic	$ 1,409,000,000
Zonegran capsules	zonisamide	Epilepsy	$ 148,000,000
		TOTAL BRAND SALES	$ 9,434,000,000
First-time generic drugs approved in 2005. Source: U.S. Food and Drug Administration.*Prior year sales are in rounded millions of dollars.			

During 2006, twenty-five first-time prescription generic drug products came on the market. These are listed alphabetically in Table VIII.

Table VIII. FIRST-TIME GENERIC PRESCRIPTION DRUGS INTRODUCED IN 2006		
Brand Name and Dosage Form	**Generic Name**	**Medical Use**
Androgel	testosterone gel	Male hormone
Climara transdermal	estradiol	Female hormone
Colestid packet	colestipol	High cholesterol
Dynacirc capsules	isradipine	Hypertension
Effexor tablets	venlafaxine	Depression
Flexeril tablets	cyclobenzaprine	Muscle relaxant
Flonase nasal spray	fluticasone propionate	Nasal congestion
Lexapro tablets	escitalopram	Depression
Metrogel topical gel	metronidazole	Antifungal
Mobic tablets	meloxicam	Arthritis pain
Norco tablets	hydrocodone/acetaminophen	Pain
Omnicef capsules; oral suspension	cefdinir	Antibiotic
Plavix tablets	clopidogrel*	Blood clots
Pravachol tablets	pravastatin	High cholesterol
Propecia tablets	finasteride	Male pattern baldness
Proscar tablets	finasteride	Enlarged prostate
Retrovir capsules	zidovudine	Viral infections (HIV)
Seasonale tablets	levonorgestrel/ethinyl estradiol	Female hormones
Surmontil capsules	trimipramine	Depression
Terazol 3 vaginal suppositories	terconazole	Fungal infections
Topamax tablets	topiramate	Convulsions
Toprol XL tablets	metoprolol succinate	Hypertension
Zaditor ophthalmic solution	ketoprofen	Eye pain

Table VIII. FIRST-TIME GENERIC PRESCRIPTION DRUGS INTRODUCED IN 2006		
Brand Name and Dosage Form	Generic Name	Medical Use
Zocor tablets	simvastatin	High cholesterol
Zoloft tablets; oral concentrate	sertraline	Depression
Source: U.S. Food and Drug Administration. *Temporary and short-lived access to marketing by Apotex.		

Generic Drug Explosion Underway

In addition to cost savings to consumers and their insurers, an explosion in generic drugs is underway for several reasons. This is largely due to both a dearth of new molecular (chemical) entities emerging from the major innovator pharmaceutical company laboratories and the large numbers of brand-name prescription drugs recently losing and anticipated to lose their patents over the next few years.

A tabulation of blockbuster and other prescription drugs expected to lose their patent protection during the next few years, have been listed earlier in Table VI in Chapter 7. The therapeutic classification, name of the manufacturer, and recent annual sales are included for each. The projected value of blockbuster prescription drugs losing patent protection in 2006 was $22 billion; in 2007, $27 billion; and in 2008, $29 billion.[64]

Biological Pharmaceuticals and Patent Expirations

Biopharmaceuticals, also referred to as biologicals, are the most expensive class of prescription drugs to date and have yet to be provided with a generics application process for timely replacement as their patents expire.

Biological pharmaceuticals are huge protein molecules of considerable complexity and present unique problems when it comes to replicating the product. These include products such as human insulin (Humulin, Lilly), alpha interferon (Intron A, Schering-Plough), erythropoietin (Procrit and

Epogen, Amgen), and filgrastim (Neupogen), all of whose patents expired prior to 2007. Unlike the others, Humulin is classified by the FDA as a drug rather than a biological. These blockbuster biopharmaceutical products, all scheduled to lose patent protection before 2007, represented total sales of $20.2 billion in 2005, according to the Generic Pharmaceutical Association.

During 2005, overall sales of biological pharmaceuticals increased 17.2 percent over the previous year, rising to $32.8 billion. This class of products is the most expensive of prescription medications and, accordingly, constitute a disproportionate share of dollars spent on pharmaceuticals. A leader at BioGenerix, a major biopharmaceuticals group, recently projected that nearly 50 percent of all new approved pharmaceuticals will be of biotechnological origin by 2010.

Biologicals or biologics or biopharmaceutical products is a broad class of therapeutic and diagnostic products isolated from a variety of natural sources, including humans, animals and microorganisms. Most biologics are comprised of highly complex mixtures difficult both to identify and characterize as compared to most other drugs which are of known chemical structure and can be synthesized in the laboratory. They may be produced by cutting-edge technologies such as biotechnology. This category of biopharmaceutical products, along with products resulting from the application of pharmacogenomics, offers the greatest hope for future prevention, treatment and cure of presently untreatable illnesses and medical conditions.

Biologicals include vaccines, blood and blood components, allergenics, somatic cells, gene therapy, tissues, and recombinant proteins. Their composition may include sugars, proteins or nucleic acids or complex combinations of these. Alternatively, they may be actual living cells or tissues. These products require special handling and techniques because they tend to be heat sensitive and are easily contaminated with microorganisms.

Examples of such products include: allergenics (allergy and contact dermatitis diagnostic patch tests); blood products (red blood cells, plasma, platelets, clotting factors and immunoglobulin); gene therapy (products to replace one's faulty or missing genetic material); human tissues and cellular products for transplantation (skin, tendons, ligaments, cartilage, human stem cells and pancreatic islets); vaccines (anthrax, chicken pox, diphtheria, hepatitis, influenza, measles, mumps, and tetanus); xenotransplantation products (live animal cells, tissues or organs) for use where human parts may not be available; and various medical devices and tests required for safeguarding blood, blood components and cellular products from infectious agents such as HIV, hepatitis, syphilis and other infections.

Biopharmaceutical products present a unique set of challenges which impede the availability of generic copies following patent expiration. Before generic biologicals, or biogenerics as they are referred to, can be made available, a safe path to replicating the complex structures and activities of the innovator product must be found. General FDA guidance on this issue, which was expected in Summer 2004, is still not available at the time of this writing. Also, at the time this writing, congressional legislation had been initiated for the purpose of expressly empowering and nudging the FDA to move ahead by taking a case-by-case approach to approving follow-on biologics.

Biologicals, such as Genotropin (somatropin, Pfizer) have been slow to move to generic or "follow on protein" status following parent expiration of the innovator biologic. This is largely due to the nature of large protein molecules and the complexity of replicating them. The FDA approval in 2006 of the "follow on" biogeneric Omnitrope (Sandoz) represents the first in this group. It appears that some legal coercion was at play in the FDA's approving this product in the absence of clear guidelines for doing so. Even so, the annual cost for a patient using this biogeneric drug was still estimated to cost about $10,000, a cost not much less than the annual cost for the original brand-name biologic. Even after biogenerics or so-called "follow-on protein" products become more commonplace, individual courses of therapy will still be considerably more expensive than for regular prescription drugs for the foreseeable future.

Pharmaceutical industry intelligence indicates that Big Pharma is beginning to focus their medicinal product research and development on biologicals or biopharmaceuticals and other high technologies, such as pharmacogenomics or application of gene therapy, as their main product

R&D focus for new drugs over the foreseeable future. While such a development holds enormous promise for finding cures for certain untreatable and incurable diseases, it does not bode well for average consumer access to affordable drugs due to the extremely higher costs for discovering, developing, manufacturing and standardizing such products. Further, potential savings from the availability of less costly follow-on generics remains but a gigantic question mark due to the absence of a procedure for approving biogenerics at the time of this writing.

Chapter 9: The Self-Care Solution with OTC Drugs and Complementary and Alternative Medicines

In this chapter we consider several new concepts. First, we discuss the concepts of *self-care* and *self-medication*, which is the most familiar aspect of self-care, a term that is more inclusive. We also discuss the two basic components of self-medication: the use of non-prescription or over-the-counter (OTC) medications, and the use of *complimentary and alternative medicines* (CAM), a concept unfamiliar to many, including some who use some CAM products regularly.

The use of OTC medications in self-care is more widespread and familiar to U.S. consumers than is the use of prescription medications, since minor illnesses amenable to self-diagnosis and self-treatment do not require the attention of a physician, and non-prescription medications are more readily available to anyone virtually everywhere a convenience store can be found. Dietary supplements—including vitamins, minerals, herbal and other botanical supplements—and homeopathic remedies are all integral components of CAM. Yes, complimentary and alternative medicine is alive and well in America, even if not to the same extent as it may be in other countries in Asia and Africa, and even though CAM is widely vilified by some as being mere quackery due to their ignorance of the breadth of coverage of CAM by definition and in practice.

Self-Care Benefits and Limitations

Self-care is a group of activities or interventions engaged in by individual consumers in order to achieve, maintain or regain a state of good health. Consumer self-care is a booming segment of overall health care in the United States. Drug smart consumers are responsible consumers who know their limits, when it comes to caring for themselves. They carefully read and take to heart accompanying labeling for the self-medication and other self-care products and programs they use. They responsibly take notice of the labeled limits on duration of symptoms before needing to seek professional attention; respect labeled cautions and warnings about product use and handling; and are intelligently cautious about self-treatment of symptoms which could represent signs of serious disease conditions.

Among the benefits of informed and responsible self-care are the following:

- reduced physician office visits;
- reduced emergency room visits;
- reduced health care costs;
- avoidance of medical errors;
- reduced absences from work;
- increased productivity;
- increased personal satisfaction;
- improved quality of life;
- increased health and drug literacy; and
- increased consumer empowerment.

In order to enjoy these benefits and safely protect their health and wellness, however, consumers must engage in responsible product and program selection and use behaviors. That is, they must understand and accept personal responsibility for their decision to manage their care:

1. It is they themselves who are responsible for knowing or learning about the types of self-care interventions available to the individual

in the absence of medical supervision, with or without the counsel of pharmacists.

2. It is they who must know or learn about lifestyles which are key to optimal and long-term health and well-being.

3. It is they who must devise and implement self-care strategies and monitor their own progress with or without the advice or counsel of a health care professional.

4. It is they who must know when their needs require the advice and/ or supervision of physicians and other health care professionals, including the pharmacist.

The author considers the following phases to be crucial for consumers to safely engage in responsible self-care:

1. *Self-awareness and adoption of healthy lifestyles* as the basic foundation for self-care activities: healthy eating; obesity/weight control; regular physical exercise; refraining from drug misuse, overuse and abuse; smoking prevention and cessation; alcohol control; responsible sexual activity; spiritual awareness and empowerment; appropriate relaxation and sleep routines, etc.

2. *Self-assessment of one's health status* and self-identification of symptoms for common, non-life-threatening and self-limiting conditions such as common allergies, athletes foot, coughs and common colds, aches and pain, headache, constipation, diarrhea, indigestion, heartburn, vaginitis, etc.

3. *Self-education* about health topics of interest and about safe and effective non-prescription drug use in a process to improve and capitalize upon one's health and drug literacy for this purpose;

4. Informed *self-selection* of safe and effective self-care products and programs.

5. *Using feedback* from self-monitoring and self-education to improve self-care and *to determine when self-care is no longer appropriate* and professional help should be sought.

These are essential steps to safe and effective consumer self-care using self-medication products such as OTC drugs, dietary supplements, including botanical and herbal supplements, and OTC homeopathic remedies. Homeopathic products are discussed briefly later in this chapter.

Self-Medication Benefits and Limitations

Self-medication is the treatment of common health problems with medications designed, labeled and approved for safe and effective use by consumers without medical supervision. It is a subset of self-care. This applies only to the use of non-prescription products. In self-medication, consumers are generally free to purchase and medicate themselves with non-prescription drugs and nutritional supplements available both in and outside of pharmacies. This practice is convenient for consumers but requires a high level of responsible behavior for self-medication to serve them well.

Consumers engaged in self-medication themselves are responsible for determining and applying appropriate self-medication interventions to provide the positive treatment outcomes they seek, including the following:

1. *Self-assessment* of one's non-prescription medication and nutritional supplement needs to maintain, regain or improve one's health status;

2. *Self-selection* of appropriate medication and nutritional and herbal supplements;

3. Self-medication with the appropriate non-prescription drug products and nutritional supplements; and

4. *Self-monitoring* of labeled dosage regimen and product effects, including intended desired effects as well as undesirable or adverse drug effects.

OTC Drugs in Self-Medication

Self-care and self-medication are worldwide health practices. Large numbers of Americans self-medicate using non-prescription (OTC) drugs, homeopathic remedies, and nutritional and herbal supplements alone or in combination with prescription drugs. In large areas of the world and, indeed, in many American communities the uninsured and underinsured are users of self-care and self-medication because of necessity. Anywhere where accessibility to medical doctors and other health care professionals is either limited or unaffordable, self-care and self-medication are necessities.

An OTC or non-prescription drug is a drug product available for use by consumers without the need for prescribing or supervision of a health care professional. On the other hand, prescription drug products can be obtained legally only on the prescription order of a licensed doctor or other legally authorized health care professional. OTC drug products are not only available through pharmacies but are also available at supermarkets, mass merchants, gas stations, etc. Prescription drug products are available only through licensed pharmacies and doctors' offices. Pharmacies may be local retail or community pharmacies, hospital pharmacies, or mail service or internet pharmacies.

The marketing and sale of non-prescription drugs are regulated by the FDA to ensure their safety for use by patients in self-care activities without the necessity of having to see a doctor to get a prescription. Many of these OTC drugs were previously in widespread use as prescription only products for extended periods of time, usually at higher doses than the doses for the over-the-counter varieties.

Consumers of OTC or non-prescription drugs purchased in excess of $15 billion of these products during 2005, according to the Consumer Health Products Association (CHPA). CHPA is the membership organization for manufacturers of OTC drugs and vitamin-mineral and other nutritional supplements. The above sales figure is based on the FDA's definition of OTC drugs and, therefore, excludes vitamins, minerals and other nutritional supplements. It also represents totals for sales in food, drug and mass merchandising stores, excluding Wal-Mart. The vast majority of OTC drugs—some 60 percent—are purchased in non-pharmacy outlets where there is no pharmacist or other health care professional to advise consumers on the selection or on the safe and effective use and handling of these products.

Non-prescription (OTC) medications are one of two major types of legal drugs available in the United States. While obtaining prescription medication requires the act of prescribing by a physician or other legally authorized health care professional, OTC drugs are available without a doctor's prescription. They are sold not only in pharmacies but also in a variety of other outlets, including supermarkets, mass merchandisers, shops in airports and hotels, gas stations, etc. In fact, more than half of OTC drugs are purchased from non-pharmacy outlets and federal authorities have refused to limit their sale to pharmacies in order to make these medicines more accessible to consumers.

In order to be sold over-the-counter as non-prescription drugs, products must have the following characteristics, according to the FDA:[65]

- their benefits must outweigh the risks associated with taking them (their margins of safety must be wide);

- their potential for misuse and abuse must be low;

- they must be amenable for use by consumers for self-diagnosed conditions;

- they can be adequately labeled; and

- health care practitioner supervision is not required for their safe and effective use.

The safe and effective use of OTC medications requires a level of maturity and health literacy to enable the user to read, understand and properly follow label directions, cautions and warnings. The self-medicating consumer must be aware of such terminology as side-effects, drug interactions and drug allergies. Responsible self-medication requires consumers to know something about the symptoms associated with the disease or conditions they intend to self-treat. It also entails focusing first on the active ingredients in each OTC product being taken in order to prevent overdosing due to duplicating doses of active drug in two or more different products.

Reading OTC product labels completely, checking out active ingredients in products at reputable Internet sites, and asking pharmacists and doctors about these products are ways for consumers to enlarge their health and drug vocabularies and elevate their health and drug literacy levels. For further information on drug and health literacy and how this affects consumer adherence to medication label directions and other instructions, see section on health literacy in Chapter 6.

Poor choices in selecting and using OTC drugs can lead to adverse effects rather than positive benefits. Prescribers may also advise patients about, recommend or prescribe non-prescription products. Health care professionals should know about all products their patients are taking—not only medication they prescribed or dispensed but also all OTC drugs, dietary supplements and herbals. Pharmacists are more readily available than other health care professionals to aid consumers in making intelligent choices of self-care products.

However, non-pharmacy outlets have neither pharmacists nor other health care personnel to answer questions about these drugs nor to advise and counsel consumers on their proper and appropriate use to insure safe and effective use. This is an especially important consideration for certain OTC drugs such as dextromethorphan and pseudoephedrine, which can be abused rather easily by teens and young adults.

Homeopathic drugs are also available in both prescription and OTC forms. OTC homeopathic drugs are discussed later in a separate section of this chapter.

Most Popular OTC Drugs

While there are many types and classifications of OTC drug products available for selection and use by consumers, the most used non-prescription drugs fall into one of the following groups:

1. Antihistamines or anti-allergy products.
2. Anti-ulcer and heartburn drug products.
3. Cough and cold medications, including antitussives or cough suppressants, decongestants and expectorants.
4. Analgesics for internal pain, headache and fever.

Estimated total sales for OTC drugs and dietary supplements in 2006, in millions of dollars, and the rankings of the top categories are as follows:[66]

- Cough and cold medicines................................$3,593.2 million
- Vitamins and supplements..............................$3,470.1 million
- Pain relievers...$2,346.0 million
- Antacids, laxatives, diarrheal remedies, etc. ....$2,127.1 million
- First aid medicines and supplies.......................$1,643.7 million
- Family planning products ....$525.0 million
- Diet products..$336.6 million
- Other remedies (consolidated) ....$5,086.5 million

These figures are based on the OTC and supplement categories as defined by ACNielsen for the 52-week period ending December 30, 2006, and reported by the CHPA. They include sales from drug and food stores, and mass merchandisers, excluding Wal-Mart.

OTC antihistamines include such drugs as loratadine (Claritin), diphenhydramine (Benadryl), brompheniramine and chlorpheniramine maleate S(Chlor-Trimeton).

Anti-ulcer drugs available OTC include H_2 antagonists and proton pump inhibitors to control hyperacidity (heartburn) in the stomach and prevent gastroesophageal reflux disease (GERD). This group includes these brands and generics: Tagamet HB200 (cimetidine), Zantac 75 (ranitidine), Pepcid AC (famotidine), Prevacid (lansoprazole), and Prilosec (omeprazole).

Cough medications with the two most popular ingredients, guaifenesin and dextromethorphan, are widely available in many configurations as liquids, tablets, and capsules as well as in some specialty dosage forms like long-acting or sustained-release products. Some of these OTC products also include antihistamines, decongestants and pain-relieving ingredients in a single product for treatment of multi-symptom coughs and colds. However, it is recommended that consumers purchase medications containing only the ingredients needed to treat the symptoms they are experiencing. This reduces their exposure to unneeded medications.

The *expectorant* guaifenesin helps make a cough more productive—as opposed to dry and harsh—by thinning mucous. *Cough suppressants* like dextromethorphan are referred to as antitussives. Products with the two ingredients combined usually carry a label designation "DM", like Robitussin DM or Tussin-DM. Products incorporating both ingredients separately and combined are widely available through pharmacies and other outlets for non-prescription drugs both as branded and generic products.

Decongestants are used to relieve congestion by reducing swelling of the mucous membranes of the nasal passages. With the recent removal of phenylpropanolamine from the market, only pseudoephedrine and phenylephrine are currently available ingredients as nasal decongestants in oral products. Only OTC products using phenylephrine as the decongestant remain as unrestricted products. Pseudoephedrine products, such as Sudafed, have been popular over the years and now must be kept under lock and key or behind a pharmacy counter in the United States. Saline (sodium chloride) sprays and drops remain available as a safer alternative for children and the elderly.

Analgesics are drugs used to relieve pain. They may also be used to reduce fever and headache. A wide variety of OTC products are available for the relief of pain, headache and fever. When used to relieve fever, they are referred to as antipyretics. However, these are basically in two categories: a) aspirin and other non-steroidal anti-inflammatories (NSAIDs), and b) acetaminophen (Tylenol). NSAIDs relieve pain by stopping the production of certain natural chemicals—prostaglandins—in the skin, muscles and joints the body. However, acetaminophen relieves pain, headache and fever by unclear mechanisms in the brain and spinal cord.

The following non-steroidal anti-inflammatories are available as brands and generics in a variety of dosage forms but principally in tablets, capsules and liquids:

1. Aspirin (Bayer and St. Joseph being the most popular brands), which is also in products like Anacin with added ingredients, like caffeine.

2. Ibuprofen, available as the major brands Advil and Motrin IB and other branded products as well as in generic form.

3. Ketoprofen, available as Orudis KT and other products as well as in generic form.

4. Naproxen, available as Aleve and other products as well as in generic form.

All of these ingredients are also available as prescription medications in higher doses and sometimes more esoteric dosage forms. Because of potentially dangerous side-effects of these drugs, the higher dose products require a doctor's prescription.

OTC Drug Approval Mechanisms

The FDA's Office of Nonprescription Products has the statutory responsibility for overseeing the approval and safe use of OTC drug products in the United States. This office provides two regulatory mechanisms for the legal marketing of OTC drug products:

1. *Filing a new drug application (NDA)* with the FDA.[67] Once the OTC product is approved, the manufacturer can only market the drug for the specific formulation (dosage form and strength) and approved labeling. This approval mechanism covers two basic subsets:

    a. filing an NDA by a sponsor for a new drug to be approved for self-medication; and

    b. an alternate route involving filing an abbreviated new drug application (ANDA) with the FDA to switch an approved prescription drug product to non-prescription status for sale over the counter.

2. *Using the FDA OTC monograph route.*[68] A drug monograph is a tool, unlike NDAs, which specify active ingredients which can be included in OTC drug products without the filing of an NDA. Such ingredients must meet standards in the applicable monograph as being "generally recognized as safe and effective" or

GRAS. These products, therefore, do not require FDA approval prior to marketing.

OTC drug products approved under both mechanisms must meet the same standards of quality in that they both must be produced in accordance with FDA's Current Good Manufacturing Practices (CGMP) standards. Active and inactive components must meet standards of the official compendia, the United States Pharmacopeia and the National Formulary, as well as approved specifications for the products themselves.

The effectiveness and safety of the OTC products are reviewed in a three-phase rule-making process which leads to the establishment of standards, as in the case of monographs, for a therapeutic category for the drug. The three phases are:

1. *First phase*—Advisory panels review the ingredients in these products to determine whether these ingredients are or could be generally recognized as safe and effective (GRASE) for use in self-medication by consumers. They also reviewed claims and recommended labeling. Labeling includes therapeutic uses, dosage, instructions for use and warnings and cautions about side-effects and misuse prevention.

2. *Second phase*—After the FDA reviews the ingredients in each class of drugs recommended by the advisory panels, public comment and other available information, the agency publishes a tentative final monograph in the *Federal Register*. Objections to the proposed final monograph can be posted and requests for hearings can be filed with the FDA Commissioner within the allotted period of time.

3. *Third phase*—This phase is the final phase of the FDA review process for OTC drugs. It involves the publication of drug monographs, which is the final regulation for the drugs. The conditions under which certain OTC drug products are recognized as GRASE are established in these monographs. Monographs are the regulatory standards for companies to use in marketing OTC

drug products which have not been approved using the NDA process.

Prescription-to-OTC Drug Switches

Many former prescription only drug products have received approval from the FDA since 1975 to be switched to OTC status for use by consumers in self-care. The first of these were antihistamines brompheniramine maleate (Dimetane, A. H. Robins) and chlorpheniramine maleate (Chlor-Trimeton, Schering) and nasal decongestants pseudoephedrine hydrochloride (Sudafed, Warner-Lambert, and Neo-Synephrine, Bayer) and pseudoephedrine sulfate (Afrinol, Schering; Duration, Plough; Dristan LA, Whitehall; and Neo-Synephrine 12 Hours), all initially approved on September 9, 1976.

Since then, many other prescription drugs have been proved safe enough for use by consumers without physician oversight, generally in lower doses than the prescription counterparts. These have included popular products in the following major classes such as:

1. Acid reducers (anti-ulcer products) cimetidine (Tagamet, SKF), famotidine (Pepcid, Merck), nazitidine (Axid, Whitehall), and omeprazole (Prilosec, Procter & Gamble).

2. The anti-allergy (antihistamine) drug loratadine (Claritin, Schering Plough), which became the first non-sedating prescription antihistamine to make the switch. Fexofenadine (Allegra, Sanofi-Aventis) and certirizine (Zyrtec, Pfizer) are among the likeliest candidates for early switches from prescription to non-prescription status.

3. Most of the non-narcotic internal or oral analgesics for pain, headache and fever relief have been switched to OTC products in lower than standard prescription doses. These include the drugs ibuprofen (Motrin, Upjohn) in 1984; naproxen (Naprosyn, Roche) in 1994; and ketoprofen (Orudis, Wyeth) in 1995.

During 2005, the FDA approved five NDAs for first-time OTC sale, with no prescription-to-OTC switches among them. Attempts to switch anti-cholesterol statins, such as Zocor (Merck's lovastatin), to OTC status in recent years have been unsuccessful as of the time of this writing. The main concern against OTC approval has been safety concerns because of adverse effects associated with this class of cholesterol medications, and a determination that physician supervision is required for the safe use of the statins. Efforts to make the switch can be expected to continue.

Consumers reap several major benefits from the switching of prescription drugs to OTC products. Switching reduces overall health care costs because:

1. It expands the range of conditions treatable with OTC products.

2. It enhances consumers' ability to engage in self-care activities.

3. It expands the most economical form of health care—self-medication with non-prescription drugs—for large segments of the population.

Using OTC Drugs with Prescription Drugs

The use of self-medication in conjunction with prescription medication poses a few additional challenges, especially with respect to possible drug interactions. So does the case of using multiple OTC drugs together. Anytime a drug substance is taken along with another drug or dietary or herbal supplement, the potential for drug interactions exists. For more information, see section below regarding important adverse drug reactions involving non-prescription medication.

Roles and Responsibilities in Self-Medication

While government regulatory agencies and the non-prescription drug industry are responsible for providing safe and effective products for consumer self-medication, consumers also play a most important role and must assume certain responsibilities in self-medication. The most basic roles and responsibilities of each of these three stakeholders are discussed below.

1. *Consumer responsibilities.* Many consumers are compelled by economic circumstances to self-medicate with non-prescription drugs. Others simply make the decision to self-medicate in order to exercise greater control over their own health care. Still others rely on non-prescription drugs for health maintenance and improvement because of the lack of ready access to health care providers for certain geographic reasons. This would be the case for people living in rural or otherwise isolated areas in the United States, as well as in large areas of many developing countries. However, all consumers using non-prescription medications must assume certain roles and responsibilities to ensure favorable health outcomes from their use as well as safety in using them. Consumers or their care-takers are responsible for the following when self-medicating:

    a. *self-diagnosing their condition* by recognizing the symptoms they are treating.

    b. *reading and understanding OTC labels* and instructions.

    c. *reading the OTC labels completely before using* or administering the OTC drugs:

        i.   the name(s) and strength(s) of the active ingredient(s);

        ii. the names of inactive or inert ingredients in the medicine;

        iii. directions for using or taking the medicine;

        iv. time limitations on the use of the medicine before professional attention should be sought;

        v.   side-effects of the drug and potential drug interactions; and

        vi. cautions, warnings, and allergy alerts.

    d. *choosing appropriate self-medication products* to treat the symptoms identified during self-diagnosis.

e.   *understanding and following labeled directions* for use as provided in the product labeling.

f.   *monitoring the effectiveness* and any problems associated with taking the OTC drug during the recommended treatment period and taking appropriate action.

g.   determining when it becomes necessary to seek professional attention.

2.   *Health care professional responsibilities.* Since self-medication is a component of self-care and since health care professionals are usually not involved in the process, the consumer is responsible for their self-medication outcomes. However, health care professionals are responsible for at least the following:

a.   consumer/patient education about the appropriateness of non-prescription medication, especially if to be used in conjunction with prescription medication.

b.   advising consumers/patients to seek medical attention from a physician or pharmacist when they are unsure of what to do; or when self-medication is inappropriate for certain conditions they may suffer from; and

c.   *pharmacists* should advise consumers/patients

i.   on the selection, proper use and handling of appropriate products for their self-diagnosed illnesses or conditions; and

ii.   when it is appropriate for them to seek medical attention.

3.   *Industry roles and responsibilities.* Manufacturers are responsible to their consumers for adherence to government regulatory codes and regulations to ensure that product integrity and the content and style of labeling, advertising and other communications are both adequate and appropriate for consumers. The result of industry's responsible citizenship will be:

a.   safe, effective and affordable products for self-medication by the average consumer;

b.   accurate, clear and responsible labeling and instructions for safe, effective use; and

c.   responsible promotion and marketing of self-medication products.

4.   *Government roles and responsibilities.* The FDA has a principal role and responsibility in ensuring that safe and effective non-prescription drug products are available for consumer use. The Office of Nonprescription Products is responsible for ensuring that:

a.   OTC drug products are properly labeled;

b.   these products are manufactured in accordance with CGMP and meet acceptable standards of quality as is the case with prescription drugs; and

c.   their benefits outweigh their risks, especially with respect to potentially harmful and long-term adverse drug effects.

The "Drug Facts" Label on OTC Drugs

Second only to the active ingredient in OTC products is the label, since these drugs are available for use without a prescription. Accordingly, as of May 2005, the mandatory change to the "Drug Facts" label was reported by FDA to be complete for all products, with few exceptions.[69] The importance of consumers reading the "Drug Facts" label cannot be overemphasized. This new OTC drug label provides detailed explanations for each label section along with a sample "Drug Facts" label. This label includes the following six "drug facts" sections and reading it is both much easier than previous OTC product labels and necessary for safe and effective use of these products:

1.   *Active ingredients.* Always the first item on the label, the generic or common name and the amounts of *the active ingredient(s) in each dosage unit* (tablet, capsule, teaspoonful, etc.) are followed by the

purpose of the medication. The active ingredient is the chemical compound that is expected to bring about the relief of symptoms being treated.

2. *Uses* or indications tell the consumer the only symptoms the medication is approved to treat.

3. *Warnings* tell the consumer other medications, foods or situations to be avoided while taking this medicine, including allergy alerts, common adverse drug reactions, especially important drug-drug, drug-disease, drug-gender or drug-age interactions or restrictions.

4. *Directions* provide the only recommended daily dosage and the frequency for taking that dosage.

5. *Other information* like conditions under which the medication should be stored such as temperature, humidity, protection from sunlight, etc., as well as other important information about the product.

6. *Inactive ingredients* contained in the product. These are chemical compounds in the dosage form which have no therapeutic effect on the body. Listed here are fillers, binders, coloring agents, preservatives, etc. Knowing these enables consumers to avoid ingredients to which they may be allergic.

Potential Problems in Using OTC Drugs

While the margin of safety for OTC drugs is much wider than for prescription drugs in general, these self-medicating products are still drugs and their use subjects the consumer to some of the same risks inherent in all drug use. Most problems associated with their use can be categorized as being a result of failures of consumers to read completely and comply with label directions, cautions, warnings or storage instructions. These potential problems include:

1. Side-effects (drowsiness, dizziness, dry mouth, blurred vision, nervousness, etc., as examples);

2.   Drug interactions with other drugs—prescription or OTC—or with alcohol, certain foods, nutritional supplements or herbs.

3.   Allergies to either one of the active ingredients or one of the inactive or inert ingredients.

4.   Inaccurate counting or measuring of doses.

5.   Improper storage of medication (high temperatures, in sunlight, in moist areas, for example).

Drug Interactions of Some Common OTC Drugs

Some important drug-drug interactions possible for the most frequently taken classes of OTC drugs, when used in conjunction with prescription drugs, are listed as illustrations only.

All of the common OTC pain relief products can interact with a number of prescription drug products to produce serious results. *Acetaminophen* interferes with the metabolism of antibiotics rifampin and isoniazid, and increases the thinning of the blood if taken along with blood thinners like warfarin. *Aspirin* intensifies the blood sugar lowering effects of several diabetes drugs like chlorpropamide and insulin; increases the blood levels of anti-seizure drugs phenytoin and valproic acid; increases blood levels of the anti-cancer drug methotrexate; and increases the blood-thinning effects of anti-clotting drugs like warfarin.

Aspirin and the other OTC NSAIDs—ibuprofen, ketoprofen and naproxen—can lead to increased blood levels of immunosuppressive drugs like cyclosporine; reduce the effectiveness of blood pressure drugs like atenolol, metoprolol and propranolol; and decrease the effectiveness of diuretics or water pills. *Ibuprofen and naproxen* can reduce the clearance of lithium by the kidneys, causing lithium concentrations to reach toxic levels in the body.

The OTC *antihistamines—brompheniramine, chlorpheniramine, dimenhydrinate, diphenhydramine and doxylamine*—interact with anti-anxiety drugs, muscle relaxants, sedatives and sleeping pills like alprazolam, diazepam, lorazepam and temazepam to increase the depressant effects on the central nervous system.

Pseudoephedrine, one of only two *decongestants* available in OTC products, interacts with monamine oxidase inhibitors (MAOIs) like phenelzine, selegiline and tranylcypromine to raise blood pressures to seriously high levels and cause problems with the heart rhythm.

The single *cough suppressant* available in OTC cough medicines, dextromethorphan, intensifies the effects of sedatives or tranquilizers when taken together.

OTC drugs will interact with the same drugs that the stronger prescription versions interact with. Because the OTC dosages are lower than the usual prescription doses, the interactions have milder effects at the recommended doses. Careful reading of OTC product labels and drug information provided with prescription medication is the safe way to go. Consumers should ask their pharmacist or physician for information on OTC drugs which can interact with their prescription medications.

Drug interactions are covered more fully in the required labeling of both the OTC drug products and the respective prescription medications. Further, more detailed information is available at FDA web sites, the Physicians Desk Reference, which includes FDA-approved labeling for prescription drug products at the time of publication; the Physicians Desk Reference for Non-Prescription Drugs; and other reputable sources like the latest Handbook of Non-Prescription Drugs published by the American Pharmacists Association. These references can be found in most libraries and major bookstores.

Complementary and Alternative Medicine

Complementary and alternative medicine encompasses products and practices outside mainstream or contemporary medicine. As defined by NCCAM, complementary and alternative medicine (CAM) is:[70]

> "a group of diverse medical and health care systems, practices, and products that are not presently considered to be part of conventional medicine."

While some still consider most, if not all, aspects of CAM to be no more than fake medicine or quackery, the National Institutes of Health has established a National Center for Complementary and Alternative Medicine (NCCAM). This agency has been charged as the Federal Government's lead agency to explore CAM practices utilizing scientific research methodology and to disseminate authoritative information to both consumers and health care professionals. This is important because large numbers of American consumers utilize a broad variety of CAM products and therapeutic practices either in self-care or as patients of regular or homeopathic physicians. As the most prominent example, the use of dietary supplements—vitamins, minerals, amino acids, and enzymes—and herbal and botanical supplements is an integral component of CAM.

For the most part, only a few of these systems, practices and products are supported by adequate scientific studies as to be recommended by health care professionals in allopathic or conventional medicine. These are the medical doctors who hold the M.D. degree, who are also referred to as allopathic physicians, and osteopathic physicians who hold the degree doctor of osteopathy (D.O.), along with nurses, pharmacists, physical therapists, psychologists, and other health care professionals who work with these physicians in conventional medical practices. Conventional medical practices involve the use of chemical and biological therapeutic agents and surgery as their primary medical interventions.

On the other hand, the practices, products and systems considered a part of CAM is ever changing, as scientific validation for the therapeutic effectiveness and safety of individual components is slowly but continually being obtained on a case by case basis.

However, according to the NCCAM, *complementary medicine may be used in conjunction with conventional medicine* as in the case of using aromatherapy following surgery. *Integrative medicine* is defined as *a combination of conventional medicine with CAM therapies* for which acceptable scientific evidence of safety and effectiveness exists. Conversely, *alternative medicine is used as a substitute for conventional medicine*; for example, using dietary therapy for colon cancer treatment rather than surgery, radiation therapy or chemotherapy.

According to NCCAM, CAM systems and interventions fall into one of the following five classifications:

- *Alternative medical systems* like *homeopathy* and *naturopathy*, which evolved before and separate from conventional medicine in the United States.

- *Biologically based systems* in which substances found in nature are used: herbs, foods and vitamins, with dietary supplements being a prime example.

- *Manipulative and body-based systems*, involving the "manipulation and/or movement of one or more parts of the body." *Chiropractic* or *osteopathic manipulation* and *massage* are prime examples of these systems.

- *Mind and body interventions* utilizing a wide variety of techniques relying on one's mind to affect body functions and symptoms. Former CAM techniques which have become part of conventional medicine include cognitive-behavioral therapy and patient support groups. Still considered to be CAM are meditation, prayer, mental healing, and art, music and dance therapy.

- *Energy therapies* involving application of energy fields, the most prominent of which is bioelectromagnetic-based therapies involving the use of electromagnetic fields.

Although used more and more in complementary medicine, acupuncture or chiropractic are not drug product based and, therefore, are not discussed further in this book. Of the CAM systems, the author limits his discussions to those involving chemical and biological interventions, as these are more commonly used by much larger segments of the population and this book is about drugs and related medicinal products. The systems to be discussed further, then, are:

- *Biologically based therapies*, more specifically the use of dietary supplements and herbal products. *Dietary supplements* include vitamins, minerals, enzymes, protein supplements, etc. *Herbs and other botanical supplements* are also included in this group.

- *Homeopathy*—Homeopathic prescription and OTC drugs, incorporating minute quantities of medicinal substances, are used to treat symptoms which would otherwise be caused by higher or more concentrated doses of the same substances.

- *Naturopathy*—Naturopathic medicine employs CAM treatments such as nutrition and lifestyle counseling, dietary supplements, medicinal plants, exercise, homeopathy and traditional Chinese medical treatments to support the body's natural healing power that establishes, maintains and restores health.

Dietary Supplements and Health

More than 100 million Americans use dietary supplements daily and more than 37.2 million consumers use herbal remedies regularly. Further, more than 16 percent of patients take prescription medications and herbal supplements concurrently.

Dietary or nutritional supplements are regulated by the FDA as foods. Yet they are regulated differently than other foods as well as differently than drugs. According to the Dietary Supplement Health and Education Act, effective in 1994, a dietary supplement:[71]

1. is intended to supplement the diet;
2. contains one or more dietary ingredients (including vitamins, minerals, herbs or other botanicals, amino acids, and other substances) or their constituents;
3. is intended to be taken by mouth as a pill, capsule, tablet or liquid; and
4. is labeled on the front panel as being a dietary supplement.

Because they are classified as foods rather than drugs, their labels can only make health claims or nutrient content claims or structure/function claims. They may not claim to diagnose, cure, mitigate, treat or prevent a disease.

Active ingredients in dietary supplements marketed in the United States prior to October 15, 1994, do not require FDA review prior to marketing due to a presumption of their being safe. Neither is proof of safety and effectiveness of dietary supplements required before marketing them in the United States.

New dietary ingredients may be marketed after giving notice to the FDA with information to substantiate the safety and effectiveness of the new ingredient. However, FDA can remove any of them from the market for ineffectiveness or safety reasons. While dietary supplements are not presently subject to Federal CGMP regulations and pre-market approval of the FDA is not required, their labeling can contain only structure-function claims.

Nutritional supplements, which are used by many along with both prescribed drugs and self-selected OTC drugs, are not approved by the FDA prior to marketing and, therefore, cannot be counted on to meet the same quality standards from manufacturer to manufacturer as is the case for both prescription and OTC drugs. However, the FDA is empowered to remove from the market those supplements found to be unsafe or ineffective.

Dietary supplements, while regulated by the FDA, are not considered to be drugs. In fact, they are defined by law as a subset of foods and regulated as such. These products contain *ingredients intended to supplement the human diet* and are available in the same variety of dosage forms that human pharmaceuticals (drugs) are available in: tablets, capsules, soft gelatin capsules, gel caps, liquids, granules and powders. They may also come as extracts or concentrates or as plant stems, roots and leaves. Dietary supplements include vitamins, minerals, herbs, other botanicals, amino acids, enzymes and other types.

Consumers must be reminded that dietary supplements can interact with both prescription and non-prescription drugs they may be taking. Dietary supplements also can cause adverse reactions in consumers depending on many factors, including, health status, quantity being taken, use of alcohol or tobacco, and the individual sensitivity and susceptibility to allergies from components of these products.

Smart consumers will note that only recently has serious attention been given to reporting adverse events from using dietary supplements. The inadequacy of the government's safety machinery regarding the use of dietary supplements is detailed in a report from the Inspector General of the Department of Health and Human Services in 2001. This report disclosed the following major findings regarding FDA's regulatory oversight of dietary supplements used in the United States:[72]

1. FDA's adverse event reporting system detects relatively few adverse events.

2. It has difficulty generating signals of possible public health concern because it lacks the information necessary to effectively analyze adverse event reports to generate them.

3. It lacks vital information to adequately assess signals of possible public health concerns generated by its adverse event reporting system primarily due to the limited availability of clinical information and of consumer use information.

4. Because of the above, FDA has rarely taken safety actions on the basis of information from its adverse event reporting system.

Vitamin supplements

Vitamins are essential substances for the proper functioning and structure of the body. They are vital to the metabolism, growth, and regulation of all body systems. Each vitamin contributes in its own unique manner and has individual recommended daily allowances. Some vitamins, like Vitamin D, are made by the body.

The most commonly used vitamins include Vitamin A; the B-complex vitamins (Vitamin B1 or thiamine, Vitamin B2 or riboflavin, Vitamin B3 or niacin, Vitamin B5 or pantothenic acid, Vitamin B6 or pyridoxine, Vitamin B7 or biotin, Vitamin B9 or folic acid, Vitamin B12 or cyanocobalamin); Vitamin C or ascorbic acid; Vitamin D or calciferol; Vitamin E or the tocopherols; Vitamin K or phylloquinone (Vitamin K_1). These vitamins are available individually, in multiple vitamin products and multiple vitamin-mineral products.

For most adults, finding and using a reputable brand multiple vitamin may be both the simplest and the safest course of action. However, all products on the market are not equal in potency, extent of release or rate of dissolution and absorption in the body. Consequently, purchasers should be vigilant in their selection of products from reputable manufacturers and suppliers. Consumers must also educate themselves on the validity of advertised claims and recommendations for specific supplements and dosages as new research information is being reported periodically and sometimes point to regressive, rather than desired, outcomes for some of the long-time hyped supplements.

Mineral supplements

Normal body functioning requires the ingestion of a delicate balance of minerals along with vitamins and other basic food components. Minerals are inorganic materials which do not contain the element carbon, as do vitamins and all living organisms, and are involved in the regulation of all body systems of living organisms. They are rather stable chemicals, some of which are required by the body in larger "recommended daily allowances" usually of 100 mg or more daily. These minerals are referred to as macro minerals. Other minerals are required in much smaller or minute quantities and are called micro or trace minerals.

Balanced diets are intended to provide all of the basic nutrients the body needs, including vitamins, minerals and proteins. However, due to a number of factors, including farming limitations imposed by soil deficiencies and other environmental influences, even "balanced diets" may be inadequate to provide the full component of both macro and trace minerals required for optimal health. Further, in today's hectic world of business and fast food, few individuals may achieve the goal of eating balanced diets, which include dairy products and fresh grains, fruits and vegetables grown in nutrient-rich soil, on a regular basis. Consequently, the need for consumers to regularly supplement their diets with vitamin, mineral and even herbal products is an intelligent option.

The macro minerals include the usual: calcium, chloride, magnesium, phosphorus, potassium, sodium and sulfur. Although we know that they are essential to healthy bodies in only minute quantities, the actual daily requirements (amounts) for these trace minerals are not known even by scientists. However, we do know that these trace elements also include boron, chromium, cobalt, copper, fluoride, iodine, iron, manganese, molybdenum, selenium and zinc.

Many consumers purchase multiple vitamin-mineral supplements to protect themselves from unhealthy dietary deficiencies. However, consumers must recognize that dietary supplements are unable to fully meet individual needs compared to eating "balanced" diets of the main food sources. Consumers with serious health problems resulting from or worsened by dietary deficiencies should consult their primary care physicians or professional dietitians.

Herbal supplements

Because there is no central fact-gathering source with input from all of the major sales channels for herbal supplements, sales estimates for these products are both incomplete and uncertain. However, based on information from the American Botanical Council, the use of herbal dietary supplements in the United States grew at double digit rates from 1995 to 1999 but have fallen to single digit figures and remained virtually flat since. Annual sales in 2000 were reported at $4.26 billion and $4.41 billion in 2005 for aggregated reporting sales channels.[73]

Beyond data provided by food, drug and mass merchandisers, the true level of herbal use in the United States is greatly underestimated because of the wide variety of channels through which these are sold and the lack of sales information from large marketers like Wal-Mart, Sam's Club, Costco, and other warehouse buying clubs. Also included among these channels are health and natural food stores, convenience stores, mail order firms, direct Internet sources, network and multi-level marketers, as well as certain health professionals like acupuncturists, chiropractors, and naturopaths.

Herbal supplements are officially dietary supplements, not drugs, and are widely used in the United States. However, the components of many herbs have pharmacological properties. Therefore, some of the most important questions one should ask before using herbals are:

1. What are they?
2. Are they effective and safe to use?

3.   How are they regulated?

4.   Are they economical to use?

While they are not regulated by the FDA as drugs and while their labels can carry no health claim or claim to cure a condition, their components can and do exert various physiological effects on the body and some can be toxic or fatal in excessive doses. Consequently, they can interact with drugs being taken by a consumer, whether they are prescription or non-prescription drugs. Herbal products are available as leaves, stems, roots and other plant parts, as well as a wide variety of dosage forms including powders, liquids, ointments, capsules and tablets.

A wide variety of herbal products is available as supplements in pharmacies and other retail outlets and their use is increasing in popularity. It is estimated that the 2006 sales of herbal products totaled some 28.8 million units compared to some 51.3 million units for mineral supplements, some 56.7 million units for non-herbal supplements, some 62.5 million units for the letter vitamins, excluding vitamin K, and some 93.5 million units of multivitamins. These estimates include sales by all pharmacies, supermarkets, and mass merchandisers, excluding Wal-Mart.[74]

Some of the most widely used herbal supplements are: aloe, cascara sagrada, cat's claw, cayenne, cinnamon, echinacea, evening primrose, garlic, ginger, gingko biloba, ginseng, glucosamine, goldenseal, grape seed extract, kava kava, melatonin, milk thistle, MSM, saw palmetto, St. John's Wort, stinging nettle, and valerian. Since most of these are plant parts, rather than individual chemicals, the composition of each can vary widely according to source or variety. These and others are available in bulk form, capsules, tablets and liquids, depending on intended use and other factors, and form the basis for homeopathic drug products, which represent infinitesimal concentrations of these ingredients, discussed below.

Effectiveness, safety and regulation of Herbals

The FDA is the federal agency responsible for monitoring the safety of food and drug products in the United States. However, the agency does not have the authority to require approval of herbals before they are marketed and sold. Since government oversight of herbal supplements is limited, consumers must carefully read herbal labels and can expect to find certain information, such as the following, on the product label:

1. Name of the herbal supplement (aloe, ginger, etc.)

2. Net quantity of contents in the container (e.g.,10 tablets)

3. Supplement facts, including serving size, amount and the active ingredient

4. Other ingredients, for which no daily values have been established, such as other herbs and amino acids

5. Name and address of manufacturer, packer or distributor

6. Sometimes a disclaimer may be required. An example of such a disclaimer is: *"This statement has not been evaluated by the Food and Drug Administration. This product is not intended to diagnose, treat, cure or prevent any disease."*

Consumers' doctors and pharmacists should always be told about any herbal supplements, as well as any other dietary supplement and OTC drug, being taken. Consumers are further responsible for and cautioned to educate themselves about any herbal supplements they are planning to take before they purchase them. It is also imperative that consumers ask their doctors or pharmacist about such products they plan on taking, especially regarding their safety, possible side-effects and potential drug interactions.

Remember that taking herbal products also carries potential health risks. For consumers with one or more serious health conditions, herbal supplements should never substitute for professional care but may supplement conventional medicine. Otherwise, their use can cause complications especially in conjunction with prescription or non-prescription drugs. Herbal products should be used along with prescription drugs only with the knowledge of one's doctor or other health care provider.

Homeopathic Drug Products

The practice of homeopathy is based on the principle that high doses of pharmacologically active substances cause symptoms when administered to healthy individuals in high doses and that the same substances, when prepared in highly diluted form, can relieve similar symptoms in health conditions which may result from different etiologies or causes. The principle involved here is referred to as the "law of similars"—like cures like.

Homeopathy is defined as an alternative medical system involving the art and science of healing the sick by using substances capable of causing the same symptoms, syndromes and conditions when administered to healthy people. Historically, homeopathy has been used in self-care by consumers not only in the United States for some 200 years but also worldwide much longer, particularly in countries where access to conventional medicine and where affordable, safe and effective drugs are limited.

The Creighton University School of Medicine, in its tutorial on homeopathy, defines the practice as a form of alternative medicine that promotes the use of infinitesimal amounts of natural substances to prevent and cure disease.

Prevalence of use of homeopathic medicines

While the regulatory process for homeopathic medicines is unfamiliar to the general public, sales for these products is a multi-million dollar business and among the top 10 best-selling non-prescription drugs in the United States in the following categories:[75]

1. Specialty analgesics.
2. Oral analgesics for children.
3. Cough-cold-flu products.

Although the sales of homeopathic remedies constitute only a fraction of the sales volumes for conventional prescription drugs and dietary supplements, a recent National Health Information Survey indicates that approximately three quarters of Americans have used CAM and some 3.6 percent have used homeopathy. One advantage of homeopathic medications is their relatively low costs compared to the retail prices for conventional drugs. Unfortunately, most homeopathic remedies are not yet covered by health insurance.

Regulation of homeopathic vs. allopathic Medicines

Like conventional or allopathic medicines, homeopathic medications are available in both prescription and non-prescription or OTC form, based on official monographs in the Homeopathic Pharmacopeia of the United States (HPUS). Their legal basis is the Federal Drug and Cosmetic Act of 1938 and they are subject to the same regulations regarding good manufacturing practices (21 CFR 210 & 211), labeling (21CFR 201) and advertising (FDA for prescriptions and FTC for non-prescription products) as for regular, allopathic drugs. A notable exception is that NDAs are not required for pre-market approval of homeopathic products.

Instead, as is the case for certain conventional OTC products, both prescription and non-prescription homeopathic drugs are subject to compliance with the official monographs of the Homeopathic Pharmacopoeia Convention of the United States (HPCUS). The HPUS has been in continuous publication since 1897.

Most homeopathic drug products in regular use are non-prescription or OTC products. Chronic use of homeopathic remedies in general is not considered consistent with the principles of this medical system. According to one major homeopathic drug manufacturer, "one of the primary tenets in homeopathy is to take as little medicine as possible until you are free of medicine."

Criteria for inclusion of a drug substance in the homeopathic pharmacopeia include the first three listed below along with at least one of the remaining four criteria:[76]

1. The HPCUS has determined that the drug is safe and effective.

2. The drug must be prepared according to specifications of the HPUS.

3. The submitted documentation must be in an approved format as set forth in the HPUS.

4. The therapeutic use of a new, non-official homeopathic drug is established by a homeopathic drug "proving" and clinical verification acceptable to the HPCUS.

5. The therapeutic use of the drug is established through published documentation that the substance was in use prior to 1962, with both subjective and any objective symptoms.

6. The therapeutic use of the drug is established by at least two adequately controlled double blind clinical studies using the drug as the single intervention, accompanied by adequate statistical analysis and acceptable subjective and, as appropriate, objective symptomatology.

7. The therapeutic use of the drug is established by one of two other forms specified by the HPCUS.

In short, the criteria for inclusion of products in the homeopathic pharmacopeia are safety and effectiveness, and conformity with the specifications of the HPUS general pharmacy section.

Although homeopathic drugs must conform to FDA CGMP regulations, their manufacturers are exempt from providing expiration dating on the product or laboratory verification of the identity and strength of each active ingredient prior to distribution.

Types of homeopathic remedies

Homeopathic remedies are available in the same basic types of common pharmaceutical dosage forms as are other types of drug products: pills (actual, old-time), tablets, capsules, solutions, tinctures, ointments and gels. Due to their nature, most homeopathic products are available in small containers. These are available in kits of various sizes, such as 50, 100 or more remedies for a wide variety of symptoms; as singles and refills for individual symptoms; or as combination products in liquid and tablet dosage forms designed to treat multiple symptoms.

For instance, one homeopathic firm sells a 200c remedy kit, a 30c remedy kit, a 30x remedy kit, a birthing remedy kit, a children's remedy kit, a top 100 remedies kit and an urgent care remedy kit.[77] It also provides liquid combinations for arthritis, congested head cold, cough, flatulence, headache, indigestion, insomnia, lumbago, menopause, mental fatigue, motion sickness and sciatica.

Combination tablet products also are available for conditions from acne, arthritis and asthma to tonsilitis, toothache and worms. Included among these are, for example, combination products for biliousness, boils, bronchitis, change of life, colds and coughs, colic, congested head colds, constipation, coughs, diarrhea, dry eyes, earache, eczema, enuresis, fever, flatulence, fungus infections, gout, hay fever, headache, hives, hoarseness, influenza, insomnia, kidney and urinary problems, mental fatigue, motion sickness, nausea and vomiting, nervous exhaustion, neuralgia, neuritis, piles, rheumatism, sciatica, sinus, sore throat, stomach disorders, teething, tonic and worms.

Also available are thousands of other products for virtually every symptom imaginable, including gels and ointments for topical application for such things as hemorrhoids and swollen veins (Hamamelis ointment); infection (Echinacea ointment); stings (Apis ointment); dry, itchy, scaly skin (sulfur 6x ointment); and many others.

Their levels of dilution, as designated on the containers or other appropriate labeling, are shown by designations such as the following:

1. "1x" refers to a dilution of active ingredient in a ratio of 1 to 9 parts inactive ingredient (diluent);

2. "1c" is a dilution of active ingredient in a ratio of 1 to 99 parts inactive ingredient (diluent).

3. A "2x" dilution is prepared using 1 part of a "1x" dilution to 9 parts of inactive ingredient (diluent).

Labels for regular (allopathic) and homeopathic prescription and nonprescription products, must include indications, and claims can include diagnosis, treatment, cure and disease prevention.

Many common homeopathic drugs are available at local pharmacies, and vary in availability according to community demand. Others can be ordered through the local pharmacy from established homeopathic product manufacturers, homeopathic pharmacies or compounding pharmacies via the Internet; from CAM practitioners; or through medical practices which include aspects of CAM along with conventional medicine. In naturopathy or naturopathic medicine, only products from natural sources are used and homeopathy is considered a component of naturopathic medicine. Naturopathy or naturopathic medicine is not discussed further in this book.

SECTION IV: Where To Go From Here

Chapter 10: User's Summary for Using Pill Secrets

The gift of life is the most precious gift known to humanity. After all, the old adage goes, self-preservation is the first law of nature. Since creation, every species known to man has lived by this first law of nature. Longevity—extended life-span—is an important component of the gift and a natural consequence of healthy living. However, for humankind, enjoying this greatest gift is much more than merely passing through the years chronologically. It is, above all, the quality of life that really matters and the hope and faith it nourishes and sustains. The responsible, rational use of prescription drugs and other remarkable tools of contemporary medicine enhance and extend the gift today in previously unimaginable ways.

The tools of high technology and properly directed ingenuity enable our scientists and health care professionals to expand and enhance this most precious gift of life in an almost infinite number of ways with physicians' healing hands, therapeutic gems, medical devices, diagnostic machinery, medical and surgical supplies, nutritional guidance, dietary and herbal supplements, and a variety of other conventional and complementary and alternative medical interventions.

So long as disease continues to afflict mankind, the search for newer and better healing gems will continue in the pharmaceutical laboratories of our industrial giants of technological innovation, the cost-cutting factories of the generic drug manufacturing competition, and among the finders, processors, and purveyors of natural substances the world over. This never ending search continues in the ongoing battle to prevent, hold at bay, and cure life-threatening infections, obliterate and control insufferable pain, and extend and improve the quality of life for all those in need. It continues for correctives for myriad diseases and unhealthy conditions resulting from genetics as well as from man's stubborn failure to adopt and follow healthy lifestyles.

Be Proactive on Issues Affecting Health

The information in this book includes reviews of a variety of issues relating to the proper use, misuse, overuse, underuse and abuse of prescription and non-prescription drugs, legal and illicit drug substances, dietary and herbal supplements and, to a lesser degree, the most commonly used aspect of alternative medicines, homeopathy.

Having read this book, the reader should now know much more about the "pills and portions" used to ameliorate ills. Most importantly, this enhanced knowledge empowers the reader to make more rational choices about the medications and supplements selected and utilized in self-care and self-medicating activities in the quest for optimal health and reduced medication and overall health care costs. Knowing more about and respecting the nature of drugs and the various pharmaceutical dosage forms in which these are made available to the consumer can now be used to advantage in guiding improved consumer compliance and the proper handling and storage of medications and related products in accordance with instructions and label directions.

Now the reader should understand better the variety of sources for and the processes involved in the discovery, production, testing, and marketing of prescription and non-prescription drugs. A simplistic understanding of the role of the major players in the nation's drug supply chain, drug industry competition, and government regulatory activities and lapses, and how these affect patient safety and out-of-pocket expenses, should now exist now that the reader is well on the way to being a drug-smart consumer.

Remember to Protect the Most Vulnerable

Because of developmental issues in the young and declining organic functions in the senior population, both groups are more prone to experience adverse drug effects. Since most prescription drugs have not been tested in infants and children, dose-related problems can be anticipated. Liver and kidney functions diminish with advancing age and significantly affect dosing of seniors due to changes in metabolism.

Ofttimes medical conditions resulting from adverse reactions to drugs among the elderly are all too often assumed to be symptoms of the normal aging process. This can result in the over-medication of seniors and further complicate already existing geriatric health problems like falls and fractures, urinary incontinence, sexual dysfunction, constipation, confusion, loss of appetite, and weight loss.

Lower levels of health and drug literacy among elderly consumers contribute to non-compliance and medication errors. In turn this leads to emergency room visits, hospitalization, increased drug and overall health care costs and, most importantly, preventable deaths.

The costs and consequences of non-compliance are heavy both in dollars and cents and in human terms. In human terms, some 125,000 deaths result annually from patients' failure to adhere to prescribed medication. Additionally, about 10 percent of hospital admissions and some 23 percent of nursing home admissions are the result of consumer failure to adhere to prescribed medication.

Annual costs to the U.S. economy for consumer literacy deficiencies are between $5 billion and $9 billion. It is already tragic that about half of all patients make medication errors and less educated consumers are much more likely to make them. Continuation of status quo is guaranteed to gradually overwhelm our health care system while leaving too many consumers at risk.

Self-Educate: Yearn to Learn for Life

Our health care professionals—doctors, dentists, pharmacists, registered nurses, etc.—are highly trained and dedicated to patient care with healthy outcomes. Governmental agencies and professional bodies aim to protect our public health by providing credentialing, licensing and other regulatory functions. As a part of the effort to ensure quality in health care, the credentialing and licensing of physicians, dentists, pharmacists, nurses and other health care professionals, require these professionals to continually update their knowledge and skills by completing specified minimum levels of continuing education in their fields.

Drug consumers, likewise, must engage in a level of "continuing education" or lifelong learning themselves on drug- and other health-related matters in order to maintain and, indeed, to expand their drug and health literacy. It is only in this fashion that consumers can adequately protect their own health and that of their families and do so in the most economical fashion. This book is intended to be a tool to assist both consumers and professionals in this effort.

Life-long learning involves many of the simple, informal activities such as reading magazine and newspaper articles regarding drug and relevant health issues, reading drug- and health-related books, listening to TV and radio programs discussing drug and other health-related issues. Drug information provided by pharmacists along with prescription medication is a great starting point, and reading this information for each drug being taken could prove to be life-saving. Health care and health professional organizations, as well as the Food and Drug Administration and the various individual institutes in the National Institutes of Health provide much valuable and reliable health information. Learning—particularly about health—is a continuous, life-long task for health smart consumers. Cease to learn and forfeit fulfilment of the wonderful gift of life.

Beware of Unreliable, Self-Serving Information

Pills and portions (medications) are readily available for treatment or cure for a great many health conditions and injuries, which is good. However, our fantasy world of health care superiority in the Western World exists largely because all too many have such an entrenched view that drugs and other medical perks are a panacea for all illnesses, discomforts and routine day-to-day challenges of human existence, whatever their nature.

In no small measure, this often seems supported by an excessive and sustained level of gullibility for accepting as gospel the hype of "un-credentialed experts" as evidenced by some of the best selling books, and other print and multimedia, as well as the Internet, all of which contributes to a virtually endless assault on the consuming public with both reliable and useful information, along with much potentially damaging misinformation. Add to this the barrage of direct-to-consumer advertising of major drug companies in the guise of consumer education, and the need to have sufficient self-education pertaining to rational drug use should be more readily apparent. In large part, it is because of the convergence of many such factors that American consumers seem programmed to rely so heavily on the ever present high tech pill for every ill.

Adopt and Promote Healthier Lifestyles

The reader's knowledge of drug actions, interactions and reactions should help one to understand the dangers in anyone's using today's miraculous healing gems as if they were mere commodities to decimate all discomfort in one's life at will. Self-contributory problems of rampart non-compliance, drug misuse and abuse, and the financial and health consequences of these should motivate drug smart consumers to rapidly mend their ways and seek ways to share the wealth.

Global problems like antibiotic resistance and microbial resistance should now carry more meaning and cause behavioral change for anyone who has been guilty of routine non-compliance with antibiotic medications and/or pressuring doctors to prescribe antibiotics for viral infections such as the common cold, influenza, most bronchitis, etc., because of being unaware of the limitations of antibiotics and serious public health problems their uninformed use can cause.

Consumer non-adherence (a.k.a. non-compliance) to drug regimens is a multi-dimensional problem which adds significantly to consumers' drug costs and the public cost of health care in the United States to the tune of some $100 billion in direct costs to the health care system annually. It also adds estimated indirect costs of some $1.5 billion in lost wages and some $50 billion in lost productivity every year.

Observations that some 19.7 million Americans 12 years or older were users of illicit drugs do not bode well for a healthy society. Further, increasing non-medical, recreational use of prescription drugs by young adults from ages 18 to 25 years of age, primarily due to an increased misappropriation and use of prescription pain relievers, is a call for immediate change.

Spending some $157 billion annually for smoking-related costs, including some $75 billion in direct medical expenses, related costs, and absenteeism and lost productivity from ill patients, should sound a deafening alarm. Estimated annual costs for alcohol abuse of some $166 billion annually, along with $18 million to $20 million annual outlays for alcohol and drug treatment, contribute to the inevitable bankruptcy of an already costly, inefficient health care system.

Consider Responsible Use of Safe Alternatives

The use of generic non-prescription drugs and dietary supplements is already a widely accepted form of self-care either mandated by economic or geographic limitations, or by choice in the United States and around the globe. The use of non-prescription—OTC—drugs is a sensible, cost-saving alternative for consumer use for self-limiting health conditions amenable to self-diagnosis, self-medication and self-monitoring. The widespread availability of OTC products beyond the pharmacy is intended to make these products conveniently available to the largest number of consumers. While their dosages, instructions for use and labeled cautions and warnings are designed to make them a safer option without the need for medical supervision, they still represent potent chemicals with the potential for substantial harm when taken improperly either alone or in conjunction with other OTC products, with prescription drugs, or with dietary or herbal supplements, as well as with certain foods. Many consumers have found dietary supplements to serve themselves well and economically. Although usually less severe and less frequent, drug interactions occur with OTC drugs and supplements the same as with prescription drugs.

Take the Dive: Be a Drug-Smart Consumer

The reader has learned a lot and should now be prepared to put this new knowledge to good use. To get started, readers should try to remember and put into practice the following basics:

1. Remember that rational drug use minimizes medication-related problems and consumer costs.

2. Be familiar with common pharmaceutical dosage forms, remembering that their basic function is as the drug delivery system for the active drug in your prescription or non-prescription drug product.

3. Remember what a drug is, its legal definition, and its basic actions, including its potential for adverse drug effects such as side-effects, drug interactions, and drug allergies.

4. Know how drug actions present both benefits and risks that are applicable to all drugs.

5. Remember why drug product labels should be read thoroughly and followed *every time* for both effective and safe treatment.

6. Take pride in having at least a basic understanding of how drugs are discovered, screened, developed, tested for effectiveness and safety, and marketed.

7. Know as much as you can about each drug you are taking.

8. Know why each drug is being taken.

9. Know what, if anything, should not be taken or eaten while taking a particular drugs.

10. Know what not to take when suffering from listed health conditions.

11. Know broadly how drug prices are determined.

12. Become more familiar with common drug-related terminology.

13. Be aware of personal allergies.

14. Know how brand-name drugs compare to generic drugs in physical properties, chemical properties, effectiveness, safety and costs.

15. Know safe alternatives available in addition to or in lieu of expensive brand-name drugs, including generic drugs, OTC drugs, and dietary and herbal supplements, based on proven safety, effectiveness and costs.

16. Remember how important issues of drug adherence/drug compliance affect drug therapy treatment outcomes and overall consumer health care costs.

17. Know what a consumer's responsibilities are for safely engaging in self-care and self-medication.

18. Find ways to share newly gained drug and health-related knowledge and skills with family members and others.

19. Always ask questions of your doctors, pharmacists and other health care professionals.

20. Yearn to learn and commit to life-long learning about disease prevention, health maintenance and improvement.

This book was organized to enable consumers and professionals to improve their drug knowledge and literacy, and to increase their awareness of important drug-related issues. In it were divulged some of the "secrets" or generally unknown information for consumers about prescription and non-prescription drugs, the dosage forms which serve as drug delivery systems, the steps involved in the discovery and marketing of new drugs, the physical chain for drug supply distribution from the manufacturer to the consumer, various drug-related problems consumers should be aware of and seek to minimize or avoid, consumer-adverse issues in the pharmaceutical industry and government regulation and, finally, cost-reduction alternatives to the more costly single-source brand-name drug products for consumers. Ultimately it is up to the individual to intelligently evaluate alternatives for possible adoption in self-care programs entered into responsibly.

Having completed this book, the reader is now ready to apply what was learned from the secrets and related background shared. It is now the reader's turn to put all of the increased drug knowledge into action toward maintaining, enhancing and improving one's own health, increasing one's personal and family safety, and making a significant dent in out-of-pocket expense for prescription and non-prescription drugs, and related health care products.

After all, my now enlightened, motivated drug consumers are empowered to read with renewed purpose and broadened understanding, and to think and use drug-related information to better manage their health care needs in partnership with their doctor, pharmacist, and other health care professionals. They now have an enhanced ability to better communicate health-related needs and preferences, to identify potential drug-related health problems, to save substantial monies in the process, and better allocate available financial resources.

As needed, readers are encouraged to revisit chapters and sections of this book periodically in order to implant the many useful ideas shared in this book firmly in memory as they embark on the journey toward optimal health as motivated drug smart consumers.

About the Author
Ira Charles Robinson

While advocating a pragmatic and cautious approach as part of a broader strategy for more rational drug therapy, Dr. Robinson encourages consumers to educate themselves in proper drug use and to evaluate safe and promising alternative therapies carefully as a means of achieving more affordable health care. He believes that increased patient/consumer education about the risks and benefits of drug use and health promotes safer, healthier lives with reduced drug and overall health care costs. He sees health literacy as a major issue in the quest to better educate consumers about safe, effective and economical prescription drug use.

In the pharmaceutical industry, the author has been a senior pharmaceutical research scientist, pharmaceutical research project leader, and technical assistant to the vice president for research at multi-national pharmaceutical industry giant, Pfizer, Inc., and president/owner of his own retail, wholesale and manufacturing operations. While at Pfizer in Brooklyn, New York, and Groton, Connecticut, he invented a novel technology for producing sustained-release drug tablets and capsules, which was later patented. He began his industrial career as a registered pharmacist in drug manufacturing at the beginning of his graduate studies at the University of Florida.

Dr. Robinson earned his B.S. in Pharmacy degree at Florida A&M University, where he was president of the student body. He also holds a Ph.D. degree in industrial pharmacy, with minors in both medicinal chemistry and industrial and systems engineering, from the University of Florida. He later served as dean and professor of pharmacy at both Florida A & M University, his undergraduate alma mater, and Howard University colleges of pharmacy.

As a registered pharmacist, he has practiced at the nation's largest pharmacy chains, Walgreens and Eckerd Drugs (now CVS), and at independently owned pharmacies in the states of Alabama, Florida, Maryland, and New York, and in Washington, D.C. At Merck-Medco (now Medco Health Solutions), then the nation's largest pharmacy benefits manager, he was a managed care pharmacist for over a decade. He also has been a patient educator for disease state self-management for consumers with diabetes and related chronic diseases.

As an international pharmaceutical health care consultant, Dr. Robinson has consulted on projects for and with the U.S. Department of State; the U.S. Agency for International Development; the Office of International Health of the U.S. Food and Drug Administration; the World Bank; the former U.S. Department of Health, Education and Welfare; the American Public Health Association; non-governmental organizations, universities, and private firms. His consultancies have taken him to Africa, the Caribbean, Europe, and the Middle East.

A native of rural Webster, Sumter County, Florida, he graduated from Mills High School. He and wife, Clarice James, have resided in Florida's Tampa Bay area for the past 20 years, after raising their four children in Washington, D.C.

Cited References

1. Federal Food, Drug and Cosmetic Act, Chapter II—Definitions, Section 201 [21 U.S.C. 321] (g)(1) (As amended through December 31, 2004), via www.fda.gov/opacom/laws/fdcact/fdcact1.htm, accessed on 4/10/07.

2. Galson, S., Report to the Nation 2005, Center for Drug Evaluation and Research, Food and Drug Administration, U.S. Department of Health and Human Services, Rockville, MD, 2006, accessed via www.fda.gov/cder/reports/rtn/2005/rtn2005.htm on 10/06/06.

3. CDER Approval Times for Priority and Standard NDAs and BLAs Calendar Years 1993–2006, via www.fda.gov/cder/rdmt/NDAapps93-06.htm, accessed 2/10/2007

4. Fast Track, Accelerated Approval and Priority Review: Accelerating Availability of New Drugs for Patients with Serious Diseases, Office of Special Health Issues, Food and Drug Administration, Rockville, MD, accessed via www.fda.gov/oashi/fast.html on 4/10/07.

5. CDER Approval Times for Priority and Standard NDAs and BLAs Calendar Years 1993–2006, via www.fda.gov/cder/NDAapps93-06.htm, accessed 2/10/2007

6. Innovation: From Laboratory to Pharmacy Shelf, Pharmaceutical Research and Manufacturers of America, Washington, DC, via http://www.phrma.org/innovation/accessed last on 4/14/2007.

7. Drug Discovery and Development: Understanding the R&D Process, Pharmaceutical Research and Manufacturers of America, Washington, DC, February 2007, via www.innovation.org/insideR&D, accessed 04/15/2007.

8. Abstracted from data provided by Verispan published in Chi, Judy, "Top Pharmaceutical Firms," *Drug Topics*, April 2, 2007, accessed 5/11/2007 via www.drugtopics.com/drugtopics/content/printContentPopup.jsp?id =414386.

9. Ibid.

10. New Blockbusters to Fuel Growth Through 2012, in Rx Annual Report 2006, Part Three: Branded Rx, Drug Store News, National Association of Chain Drug Stores, Alexandria, VA, August 28, 2006.

11. Attachment G, "The Prescription Drug Marketing Act—Report to Congress: Examination of Entities Defining Supply and Demand in Drug Distribution, Final Report: Executive Summary, 1.1 Regulatory Framework for the Distribution of Prescription Drugs," via http:// www.fda.gov/oc/pdma/report2001/attachmentg/1.html, accessed on 10/11/2006.

12. Attachment G, "The Prescription Drug Marketing Act—Report to Congress: Examination of Entities Defining Supply and Demand in Drug Distribution, Final Report: Executive Summary, 1.3 Major Categories of Wholesalers," via http://www.fda.gov/oc/pdma/ report2001/attachmentg/3.html, accessed on 10/11/2006.

13. Section 503 of the Prescription Drug Marketing Act of 1987, modified by the Prescription Drug Marketing Amendments of 1992; 21 CFR Part 203.

14. John Richardson, PBMs: The Basics and Industry Overview, The Health Strategies Consultancy, June 26, 2003, via http://www.ftc. gov/ogc/healthcarehearings/docs/030626richardson.pdf, accessed 04/09/2007.

15. Langenfeld, James, and Maness, Robert, "The Cost of 'Self-Dealing' Under the Medicare Prescription Drug Benefit," LEGG, LLC, September 9, 2003.

16. PRIME Institute, *Drug Store News*, August 30, 2001, p. 38, National Association of Chain Drug Stores, based on IMS Health data and Medicaid data from Medicaid State Drug Utilization files.

17. NACDS Industry Statistics, Total Retail Sales 2005 (Traditional Drug Stores), "Rx Sales 2005," via http://www.nacds.org, accessed on 04/09/2007.

18. Centers for Medicare and Medicaid Services, "National Health Expenditures," January 10, 2006: http://www.cms.gov/statistics/nhe.

19. NACDS Industry Statistics, "Total Retail Sales 2005 (Traditional Drug Stores): Pharmaceutical Pricing," via http://www.nacds.org, accessed on 04/09/2007.

20. NACDS Industry Statistics, "Total Retail Sales 2005 (Traditional Drug Stores): Average Estimated Retail Prescription Cost," via http://www.nacds.org, accessed on 04/09/2007.

21. Grant Thornton LLP, National Study to Determine the Cost of Dispensing Prescriptions in Community Retail Pharmacies, Prepared for the Coalition for Community Pharmacy Action: Executive Summary, Alexandria, VA, January 2007. Based on data for Medicaid prescriptions only, for six months (March–August 2006) from all 50 states, the District of Columbia and Puerto Rico.

22. Kevin Sharer, Leadership in Life Sciences: From Promise to Progress, Chairman's Address, Pharmaceutical Research and Manufacturers of America Annual Meeting, March 15, 2007, via http://www.phrma.org/about_phrma/ceo_voices/, accessed on 4/14/07.

23. Daniel Budnitz, MD, MPH, et al., National Surveillance of Emergency Department Visits for Outpatient Adverse Events, JAMA, 2006; 296:1858–1866. Accessed via http://jama.ama-assn.org/cgi/content/astract/296/15/1858, on 04/10/06.

24. The Problem of Antimicrobial Resistance, NIAID Fact Sheet, National Institute of Allergy and Infectious Disease, National Institutes of Health, U.S. Department of Health and Human Services, Bethesda, MD, April 2006, via www.niaid.nih.gov/factsheets/antimicro.htm, accessed on 11/16/06.

25. Bates DW, et al., Incidence of Adverse Drug Events and Potential Adverse Drug Events: Implications for Prevention, ADE Prevention Study Group, JAMA, 1995;274:29–34.

26. Budnitz, Daniel, et al., National Surveillance of Emergency Department Visits for Outpatient Adverse Events, JAMA, 2006; 296:1858–1866. Accessed via http://jama.ama-assn.org/cgi/content/astract/296/15/1858, on 04/10/06.

27. Flammiger, A, and Maibach, H, Dermatological Drug Dosage in the Elderly, in Medscape, 5472275,_print: Skin Therapy Letters 2006;11(8), SkinCareGuide.com. Accessed 11/29/06.

28. Bootman, JL, Cronewett, LR, et al., Preventing Medication Errors, Institute of Medicine, National Academies Press, Washington, DC, July 2006.

29. Stump, AL, Mayo, T, and Blum, A, Management of Grapefruit-Drug Interactions, American Family Physician, Vol 74, No. 4, accessed through www.aafp.org/afp/20061815/605.html.

30. National Coordinating Council on Medication Error Reporting and Prevention, About Medication Errors, via www.nccmerp.org/aboutMedErrors.html?USP_Print=true&frame=lowerfrm, accessed 08/31/2005.

31. Preventing Medication Errors, Report Brief, Institute of Medicine, National Academies Press, Washington, DC, July 2006.

32. Leape, LL, Bates, DW, Cullen, DJ, et al., Systems Analysis of Adverse Drug Events, JAMA, 1995, 274:35–43.

33. Teichman, PG, and Caffee, AE, Preventing Errors in Your Practice: Prescription Writing to Maximize Patient Safety, Family Practice Management, July/August 2002, Vol. 9, No. 7, pp. 27–30, via www.aafp.org/fpm/FPMprinter/20020700/27pres.html?print=yes, accessed 3/18/2007.

34. NCC MERP: The First Ten Years, Defining the Problem and Developing Solutions, The National Coordinating Council for Medication Error Reporting and Prevention, December 2005, p. 29.

35. Overview: Medication Adherence—Where Are We Today?, AARP, 2004, p.9.

36. Hayne, RB, "Interventions for Helping Patients to Follow Prescriptions for Medications: Cochrane Database for Systematic Review," 2001 Issue, through Frost & Sullivan White Paper: Patient Non-Adherence: Tools for Combating Persistence and Compliance Issues, 2005.

37. "Results from the 2005 Survey on Drug Use and Health: National Findings, Substance Abuse and Mental Health Services Administration, Department of Health and Human Services, Washington, DC, 2006.

38. Medscape: "Waknine, Role of Substance Abuse in Suicide and More, CDC, Morbidity and Mortality Report," November 24, 2006, via www.medscape.com/viewarticle/548290_print, accessed on November 29, 2006

39. Healthy People 2010, U.S. Department of Health and Human Services, Washington, DC, via http://www.healthypeople.gov/Document/pdf/uih/2010uih.pdf.

40. Institute Of Medicine, "Health Literacy: A Prescription to End Confusion," Report Brief, National Academies Press, Washington, DC, April 2004.

41. National Assessment of Adult Literacy, Health Literacy, Highlights of Findings, accessed on 3/14/2007 at http://nces.ed.gov/naal/indes.asp?file=assessmentof/healthliteracy/healthliteracyresearch ...

42. Divi, C, Koss, RG, et al., "Language Proficiency and Adverse Events in U.S. Hospitals: A Pilot Study," International Journal for Quality in Health Care, April 2007. 19(2):60–67.

43. Kutner, M, et al., "The Health Literacy of America's Adults: Results from the 2003 National Assessment of Adult Literacy (NCES 2006-

483), National Center for Education Statistics, U.S. Department of Education, Washington, DC, 2006.

44. Abstracted from IMS National Sales Perspectives, "Top 10 Products by U.S. Dispensed Prescriptions," IMS Health, 3/2007, via www.imshealth.com/ims/portal/front/articleC/0,2777,6599_80411799_80413615,00.html, accessed on 4/9/2007.

45. Abstracted from IMS National Sales Perspectives, "Top 10 Products by U.S. Sales, IMS Health," 3/2007, via www.imshealth.com/ims/portal/front/articleC/0,2777,6599_80408845_80411835,00.html, accessed on 4/9/2007.

46. Abstracted from IMS National Sales Perspectives, "Top 10 Therapeutic Classes by U.S. Sales," 3/2007, IMS Health, via www.imshealth.com/ims/portal/front/articleC/0,2777,6599_80408854_80411855,00.html, accessed on 4/9/2007.

47. Abstracted from IMS National Sales Perspectives, "Top 10 Therapeutic Classes by U.S. Dispensed Prescriptions," 3/2007, IMS Health, via www.imshealth.com/ims/portal/front/articleC/0,2777,6599_80411808_80413635,00.html, accessed on 4/9/2007.

48. Innovation: From Laboratory to Pharmacy Shelf, Pharmaceutical Research and Manufacturers of America, Washington, DC, via http://www.phrma.org/innovation/accessed last on 4/14/2007.

49. Source: Express Scripts 2004 Drug Trend Report via U. S. Pharmacist, June 2006.

50. Guidance to the Industry: Bioavailability and Bioequivalence Studies for Orally Administered Drug Products—General Considerations, Revision 1, Center for Drug Evaluation and Research, Food and Drug Administration, U.S. Department of Health and Human Services, Rockville, MD, March 2003, p. 3.

51. The Future of Drug Safety: Promoting and Protecting the Health of the Public, Institute of Medicine, National Academy of Sciences, Washington, D.C., September 22, 2006, via www.iom.edu.

52. "Drug Maker to Pay $430 Million in Fines, Civil Damages," *FDA Consumer Magazine*, July-August 2004, via http://www.fda.gov/fdac/features/2004/404_wl.html, accessed 10/01/2006.

53. Milton Liebman, Based on Knight Ridder analysis of Verispan's Physician Drug and Diagnostic Audit and Source Prescription Audit for the 2002–2003 12-months ending July 31, 2003. Article published in MM&M, December 2003, pp. 44–47.

54. Questions and Answers about Withdrawal of Fenfluramine (Pondimin) and Dexfenfluramine (Redux), U.S. Food and Drug Administration, Center for Drug Evaluation and Research, DHHS, via www.fda.gov/cder/news/phen/fenphenqa2.htm.

55. FDA Announces Withdrawal of Fenfluramine and Dexfenfluramine, Center for Drug Evaluation and Research, Food and Drug Administration, p. 97–32, CDER Web site, Sept 15, 1997.

56. Smith, R. Jeffrey and Birnbaum, Jeffrey H., "Drug Bill Demonstrates Lobby's Pull: Democrats Feared Industry Would Stall Bigger Changes," Washington Post, January 12, 2007, via www.washingtonpost.com, accessed 1/12/2007.

57. Press Release, Generic Drug Utilization on the Rise, Center for Medicaid and Medicare Services, U.S. Department of Health and Human Services, Atlanta, GA, Feb 8, 2007, via www.cms.hhs.gov/apps/media/press/release.aspCounter=2081.

58. Therapeutic Equivalence of Generic Drugs Letter to Health Practitioners, Center for Drug Evaluation and Research, Food and Drug Administration, Rockville, MD, January 28, 1998, via www.fda.gov/cder/news/hightgenlett.htm, accessed 3/9/07.

59. Guidance for Industry: Bioavailability and Bioequivalence Studies for Orally Administered Drug Products—General Considerations, Center for Drug Evaluation and Research, Food and Drug Administration, Rockville, MD, July 10, 2002, p. 3.

60. Ibid, p. 4.

61. Abstracted from Top 200 Generic Drugs by Retail Dollars in 2006, Verispan, VONA, via Drug Topics, February 19, 2007.

62. Abstracted from e-mail received from IMS Health, February 2006.

63. Ibid.

64. Industry Statistics: Generics in the Future, Per Bains & Co. via www.gphaonline.org, accessed on 3/3/2007.

65. Over-the-Counter (OTC) Drug Products Web Page, Office of Non-prescription Products, Food and Drug Administration, Rockville, MD, via www.fda.gov/cder/offices/otc/default.htm accessed on 2/5/2007.

66. Over-the-Counter (OTC) Drug Products Web Page, Office of Non-prescription Products, Food and Drug Administration, Rockville, MD, via www.fda.gov/cder/offices/otc/default.htm accessed on 2/5/2007

67. Office of Non-prescription Products, Regulatory Mechanisms for Marketing OTC Drug Products, OTC drug monograph regulations described in 21 CFR Part 314, via www.fda.gov/cder/offices/otc/reg_mechanisms.htm last update 11/22/2006.

68. Office of Non-prescription Products, Regulatory Mechanisms for Marketing OTC Drug Products, OTC drug monograph regulations described in 21 CFR Part 330, via www.fda.gov/cder/offices/otc/reg_mechanisms.htm last update 11/22/2006.

69. OTC Drug Facts Label, A Special Report from FDA Consumer Magazine, Food and Drug Administration, Rockville, MD, via www.fda.gov/fdac/special;testtubetopatient/otc.html, accessed on 2/5/2007.

70. CAM Basics: What Is CAM?, National Center for Complementary and Alternative Medicine, National Institutes of Health, Bethesda, MD, via http://nccam.nih.gov/health/whatiscam/, latest update on 2/12/2007.

71. Dietary Supplements: Background Information, Office of Dietary Supplements, National Institutes of Health, Bethesda, MD, via http://

ods.od.nih.gov/factsheets/dietarysupplements_pf.asp, accessed on 2/10/2007.

72. Adverse Event Reporting for Dietary Supplements: An Inadequate Safety Valve, Office of Inspector General, Department of Health and Human Services, Bethesda, MD, April 2001 (Report No. OEI-01-00-00180), via http://oig.hhs.gov/oei/reports/oei-01-00-00180.pdf accessed on 11/09/2006.

73. Blumenthal, Mark, et al., Total Sales of Herbal Supplements in the United States Show Steady Growth: Sales in Mass Market Channel Show Continued Decline, American Botanical Council, HerbalGram, 2006, 71:64–66.

74. Rob Stein, Supplement Use Doesn't Help and May Harm, Study Finds, *The Washington Post*, Washington, DC, 2/28/2007.

75. IRI Industry Report via John P. Borneman and Robert L. Field, "Regulation of homeopathic drug products," Am J Health-Syst Pharm, American Society of Health-System Pharmacists, 63:86, January 1, 2006.

76. What is the HPUS?, HPUS Online Database, Homeopathic Pharmacopoeia of the United States, via http://www.hpus.com/whatsthis.php, accessed on 1/15/2007.

77. Washington Homeopathic Products at www.homeopathyworks.com, accessed 2/28/2007.

978-0-595-43817-4
0-595-43817-2